# MAPPING AND SEQUENCING THE HUMAN GENOME

Committee on
Mapping and Sequencing the
Human Genome
Board on Basic Biology
Commission on Life Sciences
National Research Council 

NATIONAL ACADEMY PRESS
Washington, D.C. 1988

NATIONAL ACADEMY PRESS • 2101 CONSTITUTION AVENUE, NW • WASHINGTON, DC 20418

NOTICE: The project that is the subject of this report was approved by the Governing Board of the National Research Council, whose members are drawn from the councils of the National Academy of Sciences, the National Academy of Engineering, and the Institute of Medicine. The members of the committee responsible for the report were chosen for their special competences and with regard for appropriate balance.

This report has been reviewed by a group other than the authors according to procedures approved by a Report Review Committee consisting of members of the National Academy of Sciences, the National Academy of Engineering, and the Institute of Medicine.

The National Academy of Sciences is a private nonprofit, self-perpetuating society of distinguished scholars engaged in scientific and engineering research, dedicated to the furtherance of science and technology and to their use for the general welfare. Upon the authority of the charter granted to it by the Congress in 1863, the Academy has a mandate that requires it to advise the federal government on scientific and technical matters. Dr. Frank Press is president of the National Academy of Sciences.

The National Academy of Engineering was established in 1964, under the charter of the National Academy of Sciences, as a parallel organization of outstanding engineers. It is autonomous in its administration and in the selection of its members, sharing with the National Academy of Sciences the resonsibility for advising the federal government. The National Academy of Engineering also sponsors engineering programs aimed at meeting national needs, encourages education and research, and recognizes the superior achievements of engineers. Dr. Robert M. White is president of the National Academy of Engineering.

The Institute of Medicine was established in 1970 by the National Academy of Sciences to secure the services of eminent members of appropriate professions in the examination of policy matters pertaining to the health of the public. The Institute acts under the responsibility given to the National Academy of Sciences by its congressional charter to be an adviser to the federal government and, upon its own initiative, to identify issues of medical care, research, and education. Dr. Samuel O. Thier is president of the Institute of Medicine.

The National Research Council was organized by the National Academy of Sciences in 1916 to associate the broad community of science and technology with the Academy's purposes of furthering knowledge and of advising the federal government. Functioning in accordance with general policies determined by the Academy, the Council has become the principal operating agency of both the National Academy of Sciences and the National Academy of Engineering in providing services to the government, the public, and the scientific and engineering communities. It is administered jointly by both Academies and the Institute of Medicine. Dr. Frank Press and Dr. Robert M. White are chairman and vice chairman, respectively, of the National Research Council.

This study by the Board on Basic Biology was funded by the James S. McDonnell Foundation of Saint Louis, Missouri.

**Library of Congress Catalog Card Number 88-60584**

**International Standard Book Number 0-309-03840-5**



Printed in the United States of America.

First Printing, April 1988
Second Printing, January 1989

## Committee on Mapping and Sequencing the Human Genome

BRUCE M. ALBERTS (*Chairman*), University of California, San Francisco, California
DAVID BOTSTEIN, Massachusetts Institute of Technology, Cambridge, Massachusetts
SYDNEY BRENNER, MRC Unit of Molecular Genetics, Cambridge, United Kingdom
CHARLES R. CANTOR, Columbia University College of Physicians and Surgeons, New York, New York
RUSSELL F. DOOLITTLE, University of California, San Diego, California
LEROY HOOD, California Institute of Technology, Pasadena, California
VICTOR A. McKUSICK, Johns Hopkins Hospital, Baltimore, Maryland
DANIEL NATHANS, The Johns Hopkins University School of Medicine, Baltimore, Maryland
MAYNARD V. OLSON, Washington University School of Medicine, St. Louis, Missouri
STUART ORKIN, Harvard Medical School, Boston, Massachusetts
LEON E. ROSENBERG, Yale University School of Medicine, New Haven, Connecticut
FRANCIS H. RUDDLE, Yale University, New Haven, Connecticut
SHIRLEY TILGHMAN, Princeton University, Princeton, New Jersey
JOHN TOOZE, European Molecular Biology Organization, Heidelberg, Federal Republic of Germany
JAMES D. WATSON, Cold Spring Harbor Laboratory, Long Island, New York

*Advisers to the Committee*

ALBERT R. JONSEN, University of California, San Francisco, California
ERIC JUENGST, University of California, San Francisco, California

*National Research Council Staff*

JOHN E. BURRIS, *Study Director*
ROBERT A. MATHEWS, *Staff Officer* (through August 7, 1987)
CAITILIN GORDON, *Editor*
FRANCES WALTON, *Administrative Secretary*

## Commission on Life Sciences

# Preface

In the past 2 years a great deal of attention has been focused on a proposed project to map and sequence the human genome. Numerous meetings, including one sponsored by the Board on Basic Biology, have been held and a debate has developed in the biological community over the merits of such an effort. In response to questions raised by biologists about such a project, the board appointed a committee to examine the desirability and feasibility of mapping and sequencing the human genome and to suggest options for implementing the project, if it were deemed feasible.

The members of the committee are biological scientists from a variety of disciplines that deal directly or indirectly with DNA and genetic mechanisms. The committee members differ greatly in the extent of their past involvement with research on the human genome and in their potential interest in future projects to map and sequence this genome. Many of us came to this assignment with little prior knowledge of the present state of mapping and sequencing efforts. For this reason, major portions of our meetings were devoted to workshop discussions with outside experts who are deeply involved in relevant research (see Appendix C for list of speakers).

The committee asked many questions in its deliberations. Should the analysis of the human genome be left entirely to the traditionally uncoordinated, but highly successful, support systems that fund the vast majority of biomedical research? Or should a more focused and coordinated additional support system be developed that is limited to encouraging and facilitating the mapping and eventual sequencing of the human genome? If so, how can this be done without distorting the broader goals of biological research that are crucial for any understanding of the data generated in such a human genome project?

As the committee became better informed on the many relevant issues, the opinions of its members coalesced, producing a shared consensus of what should be done. This report reflects that consensus.

The committee thanks those who contributed to its work. We are grateful to all who shared their expertise with us at our meetings. In particular, we would like to thank Michael Witunski of the James S. McDonnell Foundation, which funded this study, for his insight and contributions to the process. Walter Gilbert contributed to the discussion of the issues during an initial period when he was a member of the committee. Eric Juengst and Albert Jonsen provided valuable guidance in developing and discussing the ethical and social implications of the project. The committee is indebted to the Commission on Life Sciences staff, Frances Walton, Caitilin Gordon, and Robert Mathews, whose excellent work greatly expedited the production of this report. Special thanks are due to John Burris, director of the Board on Basic Biology, for the long hours, including nights and weekends, during which he skillfully guided the report through its many drafts to a successful conclusion.

BRUCE ALBERTS, *Chairman*
Committee on Mapping and
Sequencing the Human Genome

# Contents

# MAPPING AND SEQUENCING THE HUMAN GENOME

# 1

# Executive Summary

Humans have long been intrigued by the forces that shape them and other organisms. What blueprint dictates blue eyes, brown hair, or the form of a flower? More than 100 years ago Gregor Mendel discovered that such inherited traits are controlled by cellular units that later became known as genes. In recent years, our understanding of these genes has been greatly increased by knowledge of the molecular biology of DNA, the giant molecule from which genes are formed. It is now feasible to obtain the ultimate description of genes and DNA, since recently developed techniques enable us to map (locate) the genes in the DNA of any organism and then to sequence (order) each of the DNA units, known as nucleotides, that constitute the genes.

As more of our genes are mapped and their DNA sequenced, we will have an increasingly useful resource—an essential data base that will facilitate research in biochemistry, physiology, cell biology, and medicine. This data base will have a major impact on health care and disease prevention as well as on our understanding of cells and organisms. The concept of organizing a large project to map and sequence the DNA in the genes and the intergenic regions that connect them (the entire human DNA complement or genome) has received increasing attention worldwide. Several countries have expressed interest in launching such a project. To evaluate what the United States should be doing in this area, the Board on Basic Biology of the National Research Council's Commission on Life Sciences established the Committee on Mapping and Sequencing the Human Genome, whose findings are reported in this document.

In this report the committee explores how, when, and why we should map and sequence the DNA in the human genome. In studying these issues, the committee reached the following conclusions:

- Acquiring a map, a sequence, and an increased understanding of the human genome merits a special effort that should be organized and funded specifically for this purpose. Such a special effort in the next two decades will greatly enhance progress in human biology and medicine.
- The technical problems associated with mapping and sequencing the human and other genomes are sufficiently great that a scientifically sound program require a diversified, sustained effort to improve our ability to analyze complex DNA molecules. Although the needed capabilities do not yet exist, the broad outlines of how they could be developed are clear. Prospects are therefore good that the required advanced DNA technologies would emerge from a focused effort that emphasizes pilot projects and technological development. Once established, these technologies would not only make the complete analysis of the human and other genomes feasible, but would also make major contributions to many other areas of basic biology and biotechnology.
- Important early goals of the effort should be to acquire a high-resolution genetic linkage map of the human genome, a collection of ordered DNA clones, and a series of complementary physical maps of increasing resolution. The ultimate goal would be to obtain the complete nucleotide sequence of the human genome, starting from the materials in the ordered DNA clone collection. Attaining this goal would require major (but achievable) advances in DNA handling and sequencing technologies.
- A comparative genetic approach is essential for interpreting the information in the human genome. Therefore, intensive studies of those organisms that provide particularly useful models for understanding human gene structure, function, and evolution must be carried out in parallel.
- The mapping and sequencing effort should begin primarily as a series of competing, peer-reviewed programs emphasizing technology development. Funding should include both grants to individuals and grants to medium-sized multidisciplinary groups of scientists and engineers. Because the technology required to meet most of the project's goals needs major improvement, the committee specifically recommends against establishing one or a few large sequencing centers at present.
- The human genome project should differ from present ongoing

research inasmuch as the component subprojects should have the potential to improve by 5- to 10-fold increments the scale or efficiency of mapping, sequencing, analyzing, or interpreting the information in the human genome.

- Progress toward all the above goals will require the establishment of well-funded centralized facilities, including a stock center for the cloned DNA fragments generated in the mapping and sequencing effort and a data center for the computer-based collection and distribution of large amounts of DNA sequence information. The committee suggests that the groups supplying these services be selected through open competition.

On the basis of these conclusions, the committee recommends the following:

- In view of the importance and magnitude of the task, a rapid scale-up to $200 million of additional funding per year is recommended. These additional funds should not be diverted from the current federal research budget for biomedical sciences.

A majority of the committee recommends:

- A single federal agency should serve as the lead agency for the project. This agency would receive and administer the funds for the project and would be responsible for the operation of the stock center and data center, as well as administer the peer review system utilized in determining the recipients of funds. It should work closely with a Scientific Advisory Board in developing and implementing a high standard of peer review. The Scientific Advisory Board, composed primarily of expert scientists knowledgeable in relevant fields, would provide advice not only on peer review, but also on quality control, international cooperation, coordination of efforts of the laboratories in the project, and the operations of the stock and data centers.

An outline of the major issues presented in this report follows, with genome mapping, genome sequencing, the handling of information and materials, and strategies for implementation and management of a human genome project discussed in turn.

An outline of the human genome and its central role in human biology is shown in Figure 1-1.

## GENOME MAPPING

The two main types of human genome maps are genetic linkage maps and physical maps. Genetic linkage maps are made mainly by

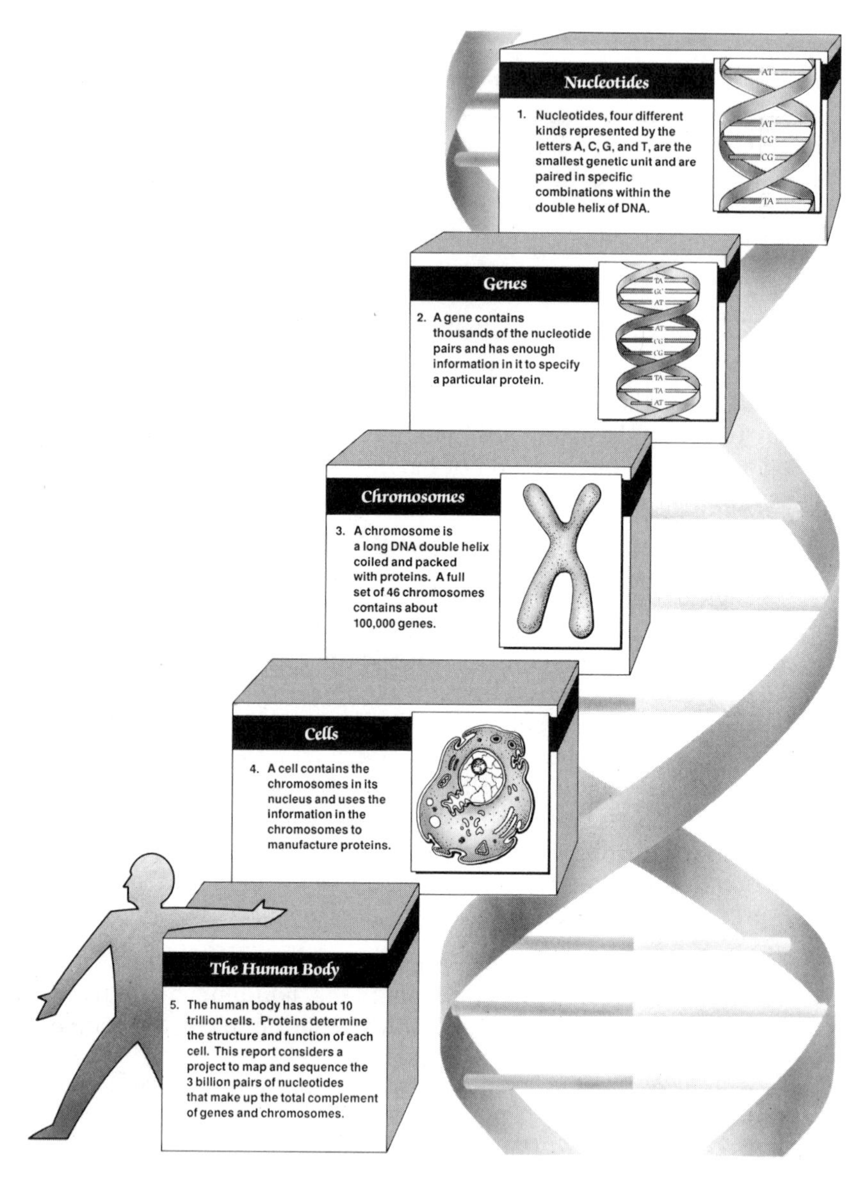


FIGURE 1-1 Adapted from an illustration by Warren Isensee for *The Chronicle of Higher Education,* September 3, 1986, with permission from the publisher.

studying families and measuring the frequency with which two different traits are inherited together, or linked. Physical maps are derived mainly from chemical measurements made on the DNA molecules that form the human genome. These maps can be of several different types and include restriction maps and ordered DNA clone collections, as well as lower resolution maps of expressed genes or anonymous (function unknown) DNA segments that are mapped by somatic cell hybridization or by in situ chromosome hybridization. All these maps share the common goal of placing information about human genes in a systematic linear order according to their relative positions along each chromosome. Knowing the location of genes and the corresponding genetic traits they produce allows us to discover patterns of genomic organization with important functional consequences and to compare humans with other mammals. Detailed maps of the human genome should quickly lead to major human health benefits. For example, by identifying genes or regions of DNA involved in several diseases, including hereditary forms of cancer, Alzheimer's disease, manic-depressive illness, Huntington's disease, and cystic fibrosis, new methods of diagnosis and treatment can be developed. Equally important, the better understanding of human biology that would follow from these studies would contribute broadly to the treatment of most diseases.

The committee believes that full-scale mapping, both genetic linkage and physical, should begin immediately. Current mapping efforts are being carried out gene by gene. Each gene is only a small part of the entire complement of DNA, and the methods involved therefore require the equivalent of repeatedly finding a needle in a haystack. In contrast, in any effort to map the entire human genome, each of the many DNA segments that are obtained by cloning the human genome will be initially kept as relevant to the project. These then represent part of a puzzle that is solved by ordering each DNA segment according to its position in the genome. The cost of obtaining any particular DNA clone in such a collection of ordered DNA clones is relatively small. As a result, a project of this type will quickly pay for itself by saving the enormous aggregate costs involved when each laboratory must find its own DNA clones.

Several recent breakthroughs in mapping methods make obtaining the type of detailed data needed in human genome maps a realistic goal. These breakthroughs range from vastly improved methods for physical mapping that rely on new techniques for separating and manipulating DNA molecules to much more accurate mathematical methods for analyzing genetic linkage data on the basis of restriction fragment length polymorphisms (RFLPs). A great deal of synergism exists between genetic linkage and physical mapping methods. Because of the simultaneous advances in both techniques, there is a real possibility that a detailed physical and genetic linkage map of the human genome could be constructed in a relatively short time. This map would be extremely useful in its own right and would set the stage for constructing the ultimate physical map—the complete DNA sequence of the human genome.

The committee concluded that the development and refinement of techniques should be emphasized early in the mapping part of the project. Despite recent advances, physical mapping methods need improvement. For example, DNA fragments as much as 10 million nucleotides long (1/300 the total human genome) can be handled only with considerable difficulty, and such large fragments cannot yet be cloned. Ordered DNA clone collections have been started, but not completed, for several organisms with genomes that are at most 1/50 the size of the human genome. Advanced technology, such as the handling of larger DNA molecules and the development of new cloning vectors for them, will expedite the preparation of such clone collections. Thus, much of the effort in the next few years should be devoted to refining existing mapping techniques and developing even more powerful new ones.

The committee believes that most support should go to groups that are attempting to map large genomes, with support for different mapping methods proceeding in parallel. Improved methods for the following would facilitate map construction and usefulness:

- Separating intact human chromosomes.
- Separating and immortalizing identified fragments of human chromosomes.
- Cloning complementary copies of expressed genes, called complementary DNAs (cDNAs), especially those that represent rare cell-, tissue-, and development-specific messenger RNAs.
- Cloning very large DNA fragments.
- Purifying very large DNA fragments, including higher resolution methods for separating such fragments.
- Ordering the adjacent DNA fragments in a DNA clone collection.
- Automating the various steps in DNA mapping, including those of DNA purification and hybridization analysis, and the development of novel methods that allow simultaneous handling of many DNA samples.

## GENOME SEQUENCING

The nucleotide sequence of the genome is the physical map at the highest level of resolution. It provides the information that constitutes the genetic complement of an individual. For the human, a total of about 3 billion ($3 \times 10^9$) nucleotides must be ordered; simply to print out such a DNA sequence would require nearly a million pages in a book like this. To obtain this critical resource in a timely fashion a special effort must be mounted, but, because of the high cost and slow rate of DNA sequencing with current technology, sequencing of the entire genome should not be initiated at present. Instead, the committee believes that two general types of effort should be encouraged to increase the efficiency of DNA sequencing.

First, pilot projects should be conducted with a goal of sequencing approximately 1 million continuous nucleotides (which is 5 to 10 times as large as the largest continuous regions that have been sequenced to date). Such projects will provide an opportunity to implement and test improvements of existing technology as they occur and will also provide a practical impetus for technological developments. They will also reveal where the most serious problems in data analysis are likely to arise in still larger projects. For example, will repetitive sequences or cloning artifacts complicate the assembly of a unique, contiguous sequence? How will new genes be identified correctly? Only by attempting relatively large-scale nucleotide sequencing will we gain insight into these issues.

Second, improvements in existing sequencing technology and the development of entirely new technologies should be vigorously encouraged. This would include applications of automation and robotics at all steps in sequencing. It is useful to think in terms of trying to achieve 5- to 10-fold incremental improvements in the scale and speed of DNA sequencing.

To derive the major benefits from a human genome sequence, it will be necessary to have an extensive data base of DNA sequences from the mouse (whose genome is the same size as that of the human) and from simpler organisms with much smaller genomes, such as bacteria, yeast, *Drosophila melanogaster* (a fruit fly), and *Caenorhabditis elegans* (a nematode worm). This information would allow the counterparts of important human genes to be readily identified in organisms in which their functional roles are generally easier to study. In addition, many genes will initially be found to be important in these other organisms and will lead to corollary human studies. Comparative sequence analysis with an organism such as the mouse is a powerful technique for distinguishing those elements of a nucleotide sequence that are important (and therefore conserved during evolution) from those that are not. To succeed, therefore, this project must not be restricted to the human genome; rather, it must include an extensive sequence analysis of the genomes of selected other species.

A mechanism of quality control is needed for the groups that are contributing large amounts of sequence information. For example, a unit could be established to redetermine a small fraction of the sequence submitted by each sequencing unit, thereby providing an independent check on the accuracy of the sequences being obtained by the unit.

## INFORMATION AND MATERIALS HANDLING

Considerable data will be generated from the mapping and sequencing project. Unless this information is effectively collected, stored, analyzed, and provided in an accessible form to the general research community worldwide, it will be of little value. This project will also require an unprecedented sharing of materials among the laboratories involved. Because access to all sequences and materials generated by these publicly funded projects should and even must be made freely available, two different types of centralized facilities will be needed: (1) information centers to collect and distribute mapping and sequencing data and (2) centers to collect and distribute materials such as DNA clones and human cell lines.

For an information center to cope effectively with the vast amount of DNA sequence data, all such data must be provided to the center in electronic or magnetic form. The information center must also be effectively linked by a computer network to all the users of the data. An initial analysis of these data should be carried out by the central facility to help in classifying the data for future research accessibility. Both at the information center and in other laboratories, extensive research in methods of sequence data analysis should be encouraged.

A facility for collecting and distributing materials should be organized to handle the cloned DNA fragments generated and mapped in the many different laboratories involved. This facility would store the appropriate DNA clones, index them according to some agreed-on plan, and then redistribute them to all laboratories that request them. The facility might also be involved in the routine conversion of large human DNA fragments, cloned as artificial chromosomes, into more readily accessible bacterial virus or cosmid DNA clone collections. It may also need to fingerprint all the DNA clones by a single method to provide a standard indexing procedure.

## IMPLEMENTATION STRATEGIES

Much of the concern that has been expressed about a project to map and sequence the human genome stems from its high projected cost and the potential changes that may result in the infrastructure of the current biological research community. The committee examined the cost of the project and concluded that an annual budget of $200 million over the next 15 years would not be excessive when compared with the value of the results that would be produced. The expenditure of $200 million per year on the project would represent roughly 3 percent of the total annual U.S. government expenditure on biological research. It would thus leave the crucial task of functional studies to traditionally supported biological research.

All decisions for funding should be based on a peer review by those expert in the methods involved. This does not mean that funding would be allocated only to individual investigators, inasmuch as multidisciplinary research centers of modest size, as well as an information center and material handling unit, will be required. Some groups may be more appropriately funded by contracts than by grants. However, the committee believes that these contracts should be awarded only after an open, peer-reviewed competition.

Genome mapping, both genetic linkage and physical, is already under way and should be intensified, although a major portion of the initial monies should be devoted to improving technologies. Large-scale sequencing should be deferred until technical improvements make this effort appropriate. This recommendation is based on the realization that the human genome is orders of magnitude larger than the genome of any other organism that has yet been mapped or sequenced. To cope with this vastly greater size, it seems advisable to establish a special competitive program that focuses on improving in 5- to 10-fold increments the scale or efficiency of mapping, sequencing, analyzing, or interpreting the biological information in the human genome.

The actual mapping of the human genome should begin now. In contrast, while a variety of pilot projects should be encouraged, only after the technology is developed and an adequate quality control procedure is established should a large-scale sequencing effort begin on the human genome.

A human genome project of this type need not threaten the existing biological research community for several reasons. First, the money ought not be provided at the expense of currently funded biological research. Second, it ought to be distributed by peer review. Third, by including selected other organisms required for the interpretation of the human genome map and sequence, the project should not mislead the public into placing a false emphasis on the uniqueness of human materials for understanding ourselves. Fourth, this project ought to include work by both small research laboratories and larger multidisciplinary centers formed by juxtaposing several small research groups having different expertise. Since individual investigators working in small groups have been the source of nearly all the major methodological breakthroughs that have driven the modern revolution in biology, the proposed organization ensures that our extraordinarily successful pattern of doing biology will be preserved.

In multidisciplinary centers, 3 to 10 research groups, each with an outstanding independent scientific director and a different but related focus, are envisioned as sharing equipment and personnel in core facilities and collaborating to accomplish a larger goal than any single group could readily achieve on its own. These centers could efficiently coordinate the large number of different experimental and computer capabilities needed for the development of techniques as well as work out optimal strategies that produce actual mapping and sequencing data.

The committee does not believe that one or a few large production centers for mapping or sequencing should be established at this time. Strong technical and intellectual advantages are obtained by distributing mapping and sequencing work among smaller multidisciplinary centers and individual research laboratories. One major advantage is that the resulting competition will stimulate research. Another is that it allows the most successful units to be identified so that the available resources can be directed to them. Moreover, the dispersal of the groups will allow close interactions to be established with a large number of other biological scientists. These interactions will be essential both for the intellectual contributions derived from other scientists and for enabling the new techniques developed in this project to be applied quickly and efficiently to a wide variety of important biological problems.

## MANAGEMENT STRATEGY

For the human genome project to be of maximum value, the committee believes that it needs to be well organized and coordinated. For this to be effectively done, a majority of the committee members feels that the project should be sited within one of three federal agencies: the National Institutes of Health, the Department of Energy, or the National Science Foundation. This lead agency would receive a specific appropriation for the project and be responsible for the disbursement of funds through a peer-review process. It would be responsible for the operation of the stock center and the data center, the coordination of the efforts of the many laboratories involved in the effort, and serve as an information clearinghouse. It would also handle the many other administrative details that will arise in a project of this magnitude.

Although the lead agency would have the ultimate responsibility for funding and policy decisions, it should draw on the advice and expertise of a Scientific Advisory Board (SAB). The SAB would be made up predominately of scientists with expertise in the methods and goals of the project. The major responsibilities of the SAB would include:

- To facilitate coordination of the efforts of the many laboratories that are expected to participate in this effort.
- To help assure the accessibility of all information and materials generated in the project by advising on the oversight of the data center and the stock center and recommending contracts where appropriate. It would oversee formation of standard terminologies and reporting formats so that the large body of information to be obtained can be readily communicated and analyzed by the entire scientific community.
- To monitor the quality of research by helping to assure a uniform standard of peer review.
- To suggest mechanisms for strict quality controls on the sequence and mapping data collected.
- To promote international cooperation, serving as a liaison to projects outside the United States regardless of their funding sources.
- To make recommendations concerning the establishment of large sequencing endeavors, thereby balancing focus with breadth.
- To publish periodic reports stating progress, problems, and recommendations for research.

The committee strongly believes that a project to map and sequence the human genome should be undertaken. It is aware of the ethical, social, and legal implications of such an effort, but feels that they can be adequately addressed. This project would greatly increase our understanding of human biology and allow rapid progress to occur in the diagnosis and ultimate control of many human diseases. As visualized, it would also lead to the development of a wide range of new DNA technologies and produce the maps and sequences of the genomes of a number of experimentally accessible organisms, providing central information that will be important for increasing our understanding of all biology.

# 2

# Introduction

All living organisms are composed of cells, each no wider than a human hair. Each of our cells contains the same complement of DNA constituting the human genome (Figure 1-1.) The DNA sequence of every person's genome is the blueprint for his or her development from a single cell to a complex, integrated organism that is composed of more than $10^{13}$ (10 million million) cells. Encoded in the DNA sequence are fundamental determinants of those mental capacities—learning, language, memory—essential to human culture. Encoded there as well are the mutations and variations that cause or increase susceptibility to many diseases responsible for much human suffering. Unprecedented advances in molecular and cellular biology, in biochemistry, in genetics, and in structural biology—occurring at an accelerating rate over the past decade—define this as a unique and opportune moment in our history: For the first time we can envision obtaining easy access to the complete sequence of the 3 billion nucleotides in human DNA and deciphering much of the information contained therein. Converging developments in recombinant DNA technology and genetics make obtaining a complete ordered DNA clone collection indexed to the human genetic linkage map a realistic immediate goal. Even determination of the complete nucleotide sequence is attainable, although ambitious. The DNA in the human genome is remarkably stable, as it must be to provide a reliable blueprint for building a new organism. For this reason, obtaining complete genetic linkage and physical maps and deciphering the sequence will provide a permanent base of knowledge concerning all human beings—a base whose utility for all activities of biology and medicine will increase with future analysis, research, and experimentation.

Even the complete sequence of DNA in the human genome will not by itself explain human biology. It will, however, serve as a great resource, an essential data bank, facilitating future research in mammalian biology and medicine. Humans, like all living organisms, are composed largely of proteins. For humans these are roughly estimated to be of 100,000 different kinds. In general, each gene codes for the production of a single protein, and a gene and its protein can be related to each other by means of the genetic code. Therefore, scientists will be able to turn to the DNA sequence of the human genome and obtain detailed information on both the structure and function of any gene or protein of interest. In addition, all genes and proteins will be classified into large family groups that provide valuable clues to their functions. In this way, many previously unknown human genes and proteins will become available for biochemical, physiological, and medical studies. The knowledge gained will have a major impact on health care and disease prevention; it will also raise challenging issues regarding rational, wise, and ethical uses of science and technology.

## GENOMES, GENES, AND GENOMIC MAPS

To understand the importance of knowledge about the human genome, one must first understand the genome's functions.

### *Genomes Consist of DNA Molecules That Contain Many Genes*

The genome of all living organisms consists of DNA, a very long two-stranded chemical polymer (Figure 2-1). Each DNA strand is composed of four different units, called nucleotides, that are linked end to end to form a long chain (Figure 2-2). These four nucleotides are symbolized as A, G, C, and T, which stand for the four bases—adenine, guanine, cytosine, and thymine—that are parts of the nucleotides. One DNA molecule, which together with some associated proteins constitutes a chromosome, differs from another in its length and in the order of its nucleotides. Each DNA molecule contains many genes, which are its functional units. These genes are arranged in a defined order along the DNA molecule. Most genes code for protein molecules—enzymes or structural elements—that determine the characteristics of a cell. In bacteria, the coding sequences of a gene are continuous strings of nucleotides, but in mammals the coding segments in a gene (called exons) are generally separated from one

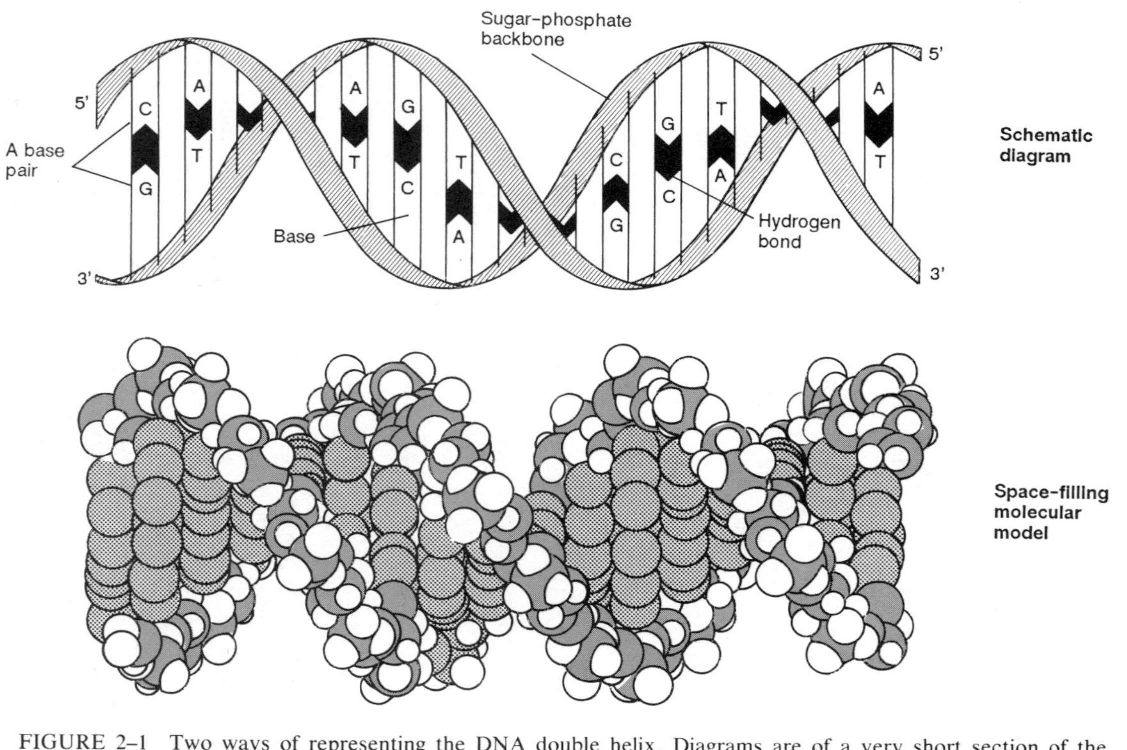


FIGURE 2–1 Two ways of representing the DNA double helix. Diagrams are of a very short section of the DNA molecule in each chromosome. The human genome contains about 200 million times the amount of DNA shown. The two strands of the DNA double helix run in opposite directions and are paired to each other by the specific fit of the complementary nucleotide pairs. Reprinted, with permission, from Alberts *et al.* (1983).

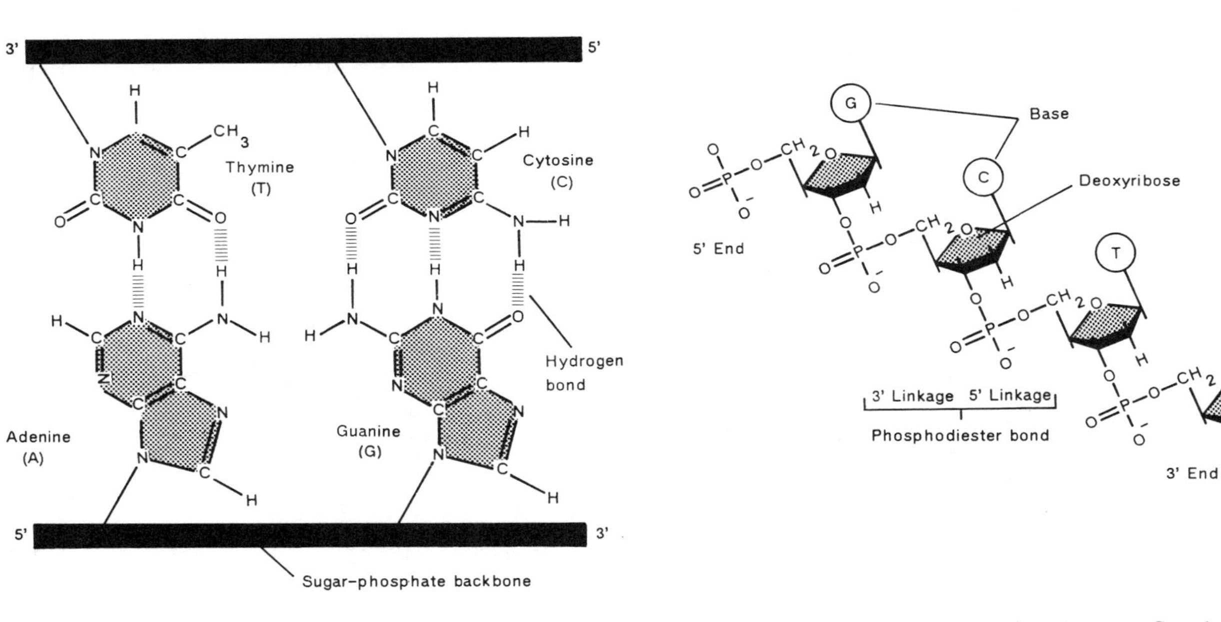


FIGURE 2–2 The nucleotides that form a DNA molecule. (A) Specific hydrogen bond interactions between G and C and between A and T bases generate complementary nucleotide pairs (that is, G always bonds with C and A always bonds with T). A haploid human genome contains 3 billion of these nucleotide pairs. (B) The chemical structure of a DNA strand. Each strand is a very long chain of the four nucleotides shown as indicated. Each nucleotide contains one of the four different bases indicated in (A) plus a sugar phosphate residue. Reprinted, with permission, from Alberts *et al.* (1983).

another by noncoding segments (called introns) (Figure 2-3). Often each exon will encode a different structural region (or domain) of a larger protein molecule. Many exons have been found to be part of a family of related coding sequences that are used in the construction of many different genes (Doolittle *et al.*, 1986). Because of the many introns in mammalian genes, a single gene is often more than than 10,000 nucleotides long, and genes that span 100,000 nucleotides are not uncommon (Table 2-1).

For the information in the coding sequences of a gene to be expressed, the DNA of a gene must first be transcribed into an RNA molecule (Figure 2-3). Before the RNA strand leaves the cell's nucleus, the intron sequences are cut out of this RNA strand by a process called RNA splicing, thereby bringing the exon sequences into contiguity. Then the RNA can be translated into a protein molecule according to the genetic code (every group of three nucleotides codes for one amino acid). Nucleotide sequences adjacent to the coding sequences in each gene encode regulatory signals for activating or inactivating transcription of the gene. Gene activity is a dynamic process; at any given time and in any given cell type, only a subset of genes is active. These active genes determine the course of embryological development and the characteristics of cells and organisms.

### *The Human Genome Is Composed of 24 Different Types of DNA Molecules*

Human DNA is packaged into physically separate units called chromosomes. Humans are diploid organisms, containing two sets of genetic information, one set inherited from the mother and one from the father. Thus, each somatic cell has 22 pairs of chromosomes called autosomes (one member of each pair from each parent) and two sex chromosomes (an X and a Y chromosome in males and two X chromosomes in females). Each chromosome contains a single very long, linear DNA molecule. In the smallest human chromosomes this DNA molecule is composed of about 50 million nucleotide pairs; the largest chromosomes contain some 250 million nucleotide pairs.

The diploid human genome is thus composed of 46 DNA molecules of 24 distinct types. Because human chromosomes exist in pairs that are almost identical, only 3 billion nucleotide pairs (the haploid genome) need to be sequenced to gain complete information concerning a representative human genome. The human genome is thus said to contain 3 billion nucleotide pairs, even though most human cells contain 6 billion nucleotide pairs.

TABLE 2-1 The Size of Some Human Genes[a]

| Gene | Gene Size (in thousands of nucleotides) | mRNA Size (in thousands of nucleotides) | Number of Introns |
|---|---|---|---|
| *Small* | | | |
| Alpha globin | 0.8 | 0.5 | 2 |
| Beta globin | 1.5 | 0.6 | 2 |
| Insulin | 1.7 | 0.4 | 2 |
| Apolipoprotein E | 3.6 | 1.2 | 3 |
| Parathyroid | 4.2 | 1.0 | 2 |
| Protein kinase C | 11 | 1.4 | 7 |
| *Medium* | | | |
| Collagen I | | | |
| Pro-alpha–1(I) | 18 | 5 | 50 |
| Pro-alpha–2(I) | 38 | 5 | 50 |
| Albumin | 25 | 2.1 | 14 |
| High-mobility group CoA reductase | 25 | 4.2 | 19 |
| Adenosine deaminase | 32 | 1.5 | 11 |
| Factor IX | 34 | 2.8 | 7 |
| Catalase | 34 | 1.6 | 12 |
| Low-density lipoprotein receptor | 45 | 5.5 | 17 |
| *Large* | | | |
| Phenylalanine hydroxylase | 90 | 2.4 | 12 |
| Factor VIII | 186 | 9 | 25 |
| Thyroglobulin | 300 | 8.7 | 36 |
| *Very large* | | | |
| Duchenne muscular dystrophy | >2,000 | ~17 | ~50 |

[a] Table provided by Victor McKusick.

DNA is a double helix: Each nucleotide on a strand of DNA has a complementary nucleotide on the other strand. The information on one DNA strand is therefore redundant to that on the other (that is because of complementary base pairing (Figure 2-2A), one can in principle determine the nucleotide sequence of one strand from the other). However, it is currently necessary to determine the sequences of the nucleotides on the two DNA strands separately to achieve the

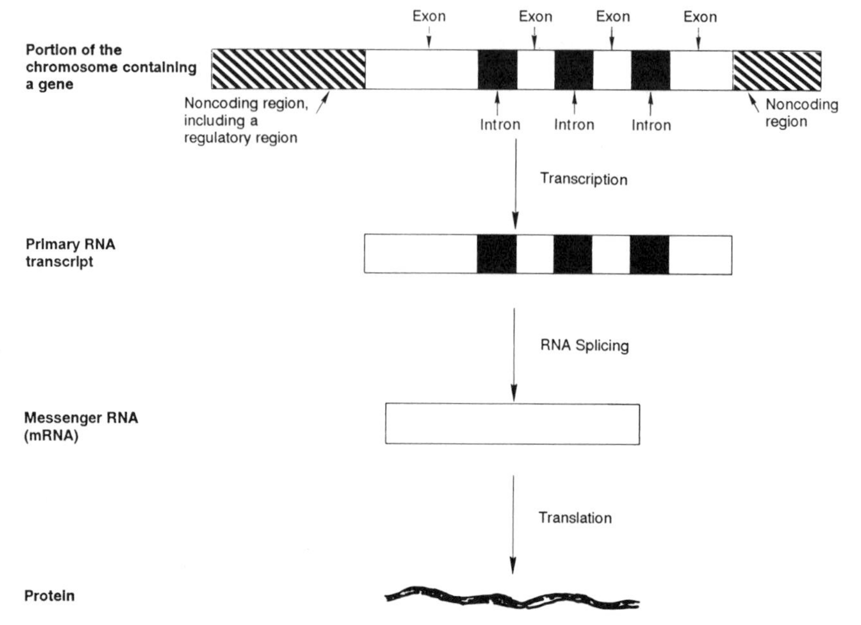


FIGURE 2–3 How genes are expressed in human cells. Each gene can specify the synthesis of a particular protein. Whether a gene is off or on depends on signals that act on the regulatory region of the gene. When the gene is on, the entire gene is transcribed into a large RNA molecule (primary RNA transcript). This RNA molecule carries the same genetic information as the region of DNA from which it is transcribed because its sequence of nucleotides is determined by complementary nucleotide pairing to the DNA during RNA synthesis. The RNA quickly undergoes a reaction called RNA splicing that removes all of its intron sequences and joins together its coding sequences (its exons). This produces a messenger RNA (mRNA) molecule. The RNA chain is then used to direct the sequence of a protein (translation) according to the genetic code in which every three nucleotides (a codon) specifies one subunit (an amino acid) in the protein chain.

desired accuracy of any DNA sequence, with the sequence of each strand being used as a check on the other. For this reason, a total of 6 billion nucleotides must actually be sequenced to order the 3 billion nucleotide pairs in the haploid human genome.

The average size of a protein molecule allows one to predict that there are approximately 1,000 nucleotide pairs of coding sequence per gene. Since humans are thought to have about 100,000 genes, a total of about 100 million nucleotide pairs of coding DNA must be present in the human genome. That this is only about 3 percent of the total size of the genome leads one to conclude that less than 5 percent of the human genome codes for proteins. The vast bulk of human DNA lies between genes and in the introns. Some of the noncoding DNA plays a role in regulating gene activity, while other portions are believed to be important for organizing the DNA into chromosomes and for chromosome replication (Alberts *et al.*, 1983; Lewin, 1987). The function of most noncoding regions of the human genome, however, is unknown; much of this DNA may have no function at all.

### *The Human Genome Can Be Mapped in Many Different Ways*

It would be enormously useful to determine the order and spacing of all the genes that make up the genome. Such information is said to constitute a gene or genome map. Since there are 24 different DNA molecules in the human genome, a complete human gene map consists of 24 maps, each in the linear form of the DNA molecule itself.

One type of useful genome map is the messenger RNA (mRNA) or exon map. Cellular enzymes transcribe or copy all of an organism's genes into mRNAs so that the functions of the genes can be expressed. Complementary DNA (cDNA) of all the mRNAs present in an organism can be synthesized enzymatically with reverse transcriptase. These cDNAs can then be cloned and used to locate the corresponding genes on a chromosome map. In this way, the genes can be mapped in the absence of knowledge of their function. Another type of genome map would consist of an ordered set of the overlapping DNA clones that constitute an entire chromosome. Both the exon map and the ordered set of DNA clones are usually referred to as physical maps. Alternatively, the position of a gene can be mapped by following the effect of the expression of the gene on the cells containing it. Here, a map is constructed on the basis of the frequencies of coinheritance of two or more genetic markers. This type of map is referred to as a genetic linkage map. The distinction between physical and genetic linkage maps is discussed in detail in Chapter 4.

Maps of the human genome can be made at many different scales, or levels of resolution. Low-resolution physical maps have been derived from the distinctive patterns of bands that are observed along each chromosome by light microscopy of stained chromosomes. Genes have been physically associated with particular bands or clusters of bands in a number of ways. These associations permit genes to be mapped only approximately since a given gene might be assigned to a region of about 10 million nucleotides containing several hundred genes. Its exact position on the chromosome must be determined by more precise methods.

Maps of higher resolution are based on sites in the DNA cut by special proteins called restriction enzymes. Each enzyme recognizes a specific short sequence of four to eight nucleotides (a restriction site) and cuts the DNA chain at one point within the sequence (Watson *et al.*, 1983). Since dozens of different sequences are recognized by one or another enzyme, and these sequences are closely spaced throughout the genome, high-resolution physical maps can be constructed by determining the relative location of different restriction sites precisely. Of particular value in human gene mapping are restriction sites that are highly variable (or polymorphic) in the population. DNA lacking a specific restriction site yields a larger restriction fragment when cut by the enzyme than DNA containing the site; hence, the designation restriction fragment length polymorphism (RFLP). Hundreds of polymorphic restriction sites have so far been identified and mapped in the human genome. Some disease-related genes have already been localized by determining the frequency of coinheritance of RFLPs and genetic diseases (Gusella *et al.*, 1983). Examples of these diseases include cystic fibrosis, Duchenne muscular dystrophy, Alzheimer's disease, and neurofibromatosis. Identifying a much larger number of useful polymorphic restriction sites should make it possible to map disease-related genes precisely enough to greatly facilitate the isolation of any human gene.

The map based on a collection of ordered clones of genomic fragments has a special value (see Chapter 4). In such a map, not only are the genomic positions of restriction fragments known, but each fragment is available as a clone that can be propagated and distributed to interested researchers. Such clones are immensely valuable because they serve as the starting point for gene isolation, for functional analyses, and for the determination of nucleotide sequences.

The ultimate, highest resolution map of the human genome is the nucleotide sequence, in which the identity and location of each of 3 billion nucleotide pairs is known (see Chapter 5). Only such a sequence reveals all or nearly all the information in the human genome. A number of specific regions of human DNA have already been analyzed in this way, providing information about the structure of genes and their encoded proteins in both normal and abnormal individuals and about sequences that regulate gene expression (Figure 2-4). At present, however, the nucleotide sequence of substantially less than 0.1 percent of the human genome is known. This includes the sequence containing 0.5 percent of our genes.

CCCTGTGGAGCCACACCCTAGGGTTGGCCA
ATCTACTCCCAGGAGCAGGGAGGGCAGGAG
CCAGGGCTGGGCATAAAAGTCAGGGCAGAG
CCATCTATTGCTTACATTTGCTTCTGACAC
AACTGTGTTCACTAGCAACTCAAACAGACA
CCATGGTGCACCTGACTCCTGAGGAGAAGT
CTGCCGTTACTGCCCTGTGGGGCAAGGTGA
ACGTGGATGAAGTTGGTGGTGAGGCCCTGG
GCAGGTTGGTATCAAGGTTACAAGACAGGT
TTAAGGAGACCAATAGAAACTGGGCATGTG
GAGACAGAGAAGACTCTTGGGTTTCTGATA
GGCACTGACTCTCTCTGCCTATTGGTCTAT
TTTCCCACCCTTAGGCTGCTGGTGGTCTAC
CCTTGGACCCAGAGGTTCTTTGAGTCCTTT
GGGGATCTGTCCACTCCTGATGCTGTTATG
GGCAACCCTAAGGTGAAGGCTCATGGCAAG
AAAGTGCTCGGTGCCTTTAGTGATGGCCTG
GCTCACCTGGACAACCTCAAGGGCACCTTT
GCCACACTGAGTGAGCTGCACTGTGACAAG
CTGCACGTGGATCCTGAGAACTTCAGGGTG
AGTCTATGGGACCCTTGATGTTTTCTTTCC
CCTTCTTTTCTATGGTTAAGTTCATGTCAT
AGGAAGGGGAGAAGTAACAGGGTACAGTTT
AGAATGGGAAACAGACGAATGATTGCATCA
GTGTGGAAGTCTCAGGATCGTTTTAGTTTC
TTTTATTTGCTGTTCATAACAATTGTTTTC
TTTTGTTTAATTCTTGCTTTCTTTTTTTTT
CTTCTCCGCAATTTTTACTATTATACTTAA
TGCCTTAACATTGTGTATAACAAAAGGAAA
TATCTCTGAGATACATTAAGTAACTTAAAA
AAAAACTTTACACAGTCTGCCTAGTACATT
ACTATTTGGAATATATGTGTGCTTATTTGC
ATATTCATAATCTCCCTACTTTATTTTCTT
TTATTTTTAATTGATACATAATCATTATAC
ATATTTATGGGTTAAAGTGTAATGTTTTAA
TATGTGTACACATATTGACCAAATCAGGGT
AATTTTGCATTTGTAATTTTAAAAAATGCT
TTCTTCTTTTAATATACTTTTTTGTTTATC
TTATTTCTAATACTTTCCCTAATCTCTTTC
TTTCAGGGCAATAATGATACAATGTATCAT
GCCTCTTTGCACCATTCTAAAGAATAACAG
TGATAATTTCTGGGTTAAGGCAATAGCAAT
ATTTCTGCATATAAATATTTCTGCATATAA
ATTGTAACTGATGTAAGAGGTTTCATATTG
CTAATAGCAGCTACAATCCAGCTACCATTC
TGCTTTTATTTTATGGTTGGGATAAGGCTG
GATTATTCTGAGTCCAAGCTAGGCCCTTTT
GCTAATCATGTTCATACCTCTTATCTTCCT
CCCACAGCTCCTGGGCAACGTGCTGGTCTG
TGTGCTGGCCCATCACTTTGGCAAAGAATT
CACCCCACCAGTGCAGGCTGCCTATCAGAA
AGTGGTGGCTGGTGTGGCTAATGCCCTGGC
CCACAAGTATCACTAAGCTCGCTTTCTTGC
TGTCCAATTTCTATTAAAGGTTCCTTTGTT
CCCTAAGTCCAACTACTAAACTGGGGGATA
TTATGAAGGGCCTTGAGCATCTGGATTCTG
CCTAATAAAAAACATTTATTTTCATTGCAA
TGATGTATTTAAATTATTTCTGAATATTTT
ACTAAAAAGGGAATGTGGGAGGTCAGTGCA
TTTAAAACATAAAGAAATGATGAGCTGTTC
AAACCTTGGGAAAATACACTATATCTTAAA
CTCCATGAAAGAAGGTGAGGCTGCAACCAG
CTAATGCACATTGGCAACAGCCCCTGATGC
CTATGCCTTATTCATCCCTCAGAAAAGGAT
TCTTGTAGAGGCTTGATTTGCAGGTTAAAG
TTTTGCTATGCTGTATTTTACATTACTTAT
TGTTTTAGCTGTCCTCATGAATGTCTTTTC

FIGURE 2-4 The DNA sequence of the human gene for beta-globin (a protein of 146 amino acids that forms part of the hemoglobin molecule that carries oxygen in the blood). The sequence of only one of the two DNA strands is given since the other one has a precisely complementary sequence. The sequence should be read from left to right in successive lines down the page, as if it were normal text. The human genome is about 2 million times as long as this small gene of 1,500 nucleotides (see Table 2–1), which contains three exons and two introns (the boundaries between exons and introns are not indicated here). Reprinted, with permission, from Alberts *et al.* (1989).

## MEDICAL IMPLICATIONS OF DETAILED HUMAN GENOME MAPS

Advances in molecular genetics made over the past two decades are already having a major impact on medical research and clinical care. The ability to clone and analyze individual genes and to deduce the amino acid sequences of encoded proteins has greatly increased our understanding of genetic disorders, the immune system, endocrine abnormalities, coronary artery disease, infectious diseases, and cancer. A few proteins produced on a commercial scale by recombinant DNA methods are available for therapeutic use or in clinical trials, and many more are in earlier developmental stages. Recent progress in determining the genetic basis for such neurological and behavioral disorders as Huntington's disease (Gusella *et al.*, 1983), Alzheimer's disease (St George-Hyslop *et al.*, 1987), and manic-depressive illness (Egeland *et al.*, 1987) promises new insights into these common and serious conditions. Higher resolution maps of the human genome will accelerate progress in understanding disease pathogenesis and in developing new approaches to diagnosis, treatment, and prevention in many areas of medicine. In Chapter 3 the potential medical impact of a detailed human genomic map is discussed further.

## IMPLICATIONS FOR BASIC BIOLOGY

The generation of a physical map of the human genome and the determination of its nucleotide sequence will provide an important research tool for basic biology. This is especially true because we expect a human genome project to support mapping and sequencing investigations that are carried out concurrently in other extensively studied organisms, including the *Escherichia coli* bacterium, the lower eucaryote *Saccharomyces cerevisiae* (a yeast), the nematode worm *Caenorhabditis elegans*, the fruit fly *Drosophila melanogaster*, the mouse *Mus musculus*, and possibly also a plant such as maize or *Arabidopsis*. Analyzing these genomes will approximately double the total amount of DNA to be mapped and sequenced. But the additional effort will make it possible to test the function of genes that have been identified in humans in other organisms that are experimentally accessible and for which powerful genetic techniques exist. It will thereby be possible to firmly establish the exact role of these genes in important biological processes. Conversely, proteins that are discovered to be of special interest in any of these other organisms can be immediately identified by amino acid homology in the human, thereby enabling investigators to conduct well-focused studies of the function of the corresponding human protein and its gene. The extensive DNA sequence and functional comparisons that are generated will also represent an invaluable resource for evolutionary biologists. These and other implications for basic biology are discussed in greater detail in Chapter 3.

## EXPECTED TECHNOLOGICAL DEVELOPMENTS GENERATED BY A HUMAN GENOME PROJECT AND THEIR IMPACT ON BIOLOGICAL RESEARCH

The process of mapping and sequencing the human genome is likely to have important spin-offs in the form of new technologies with broad applicability in both basic and applied biological research. For example, efficient methods for mapping complex genomes are still being developed, and a human genome project would accelerate this process. Such methods include improvements in the production, separation, and cloning of large pieces of DNA and methods for constructing an ordered set of genomic clones (see Chapter 4). This methodology will be directly applicable to the development of a physical map of the genomes of many experimentally and commercially important animals and plants.

Similarly, an effort to sequence the human genome will require much more efficient nucleotide sequencing technology than now exists (see Chapter 5). These improvements will greatly reduce the time spent on DNA sequencing in individual research laboratories. In the future, the development of institution-wide or regional sequencing facilities equipped with highly automated instruments could serve a large number of scientists, freeing them to concentrate on more advanced stages of their research problems.

Finally, the generation of a detailed map of the human genome will require new computer-based methods for collecting, storing, and analyzing the large amount of information expected (see Chapter 6). These methods can easily be adapted to handling analogous data from other organisms. Scientists will thus have immediately available through computer networks an enormous store of biological information supported by methods for using it, such as clone collections; these resources are likely to have a major beneficial impact on the way that individual scientists do research.

## IMPACT ON THE RESEARCH BY SMALL GROUPS

One of the key features and attractions of biomedical research today is that it is based primarily on the efforts of small, independent groups of scientists. The major advances of the past decades can be traced to the creativity of these groups, or even to single individuals, often near the beginnings of their careers. Mapping and sequencing the human genome, on the other hand, is likely to require organizational arrangements on a considerably larger scale than is customary in other biological research. Some see this as a threat to the independence of individual investigators. In the committee's view, however, a mapping and sequencing project should have as its primary goal an increase in the power and range of the research potential of small groups of individuals.

The complete nucleotide sequences of the genomes of the several organisms of major experimental interest will provide a critical reference data base for interpreting and studying the many human genes that will be discovered. To take just one example, an individual cancer researcher who discovers a new oncogene in a human tumor will have immediate access by computer search to all the proteins that are likely to have a related function in lower organisms. Since these genes can be experimentally manipulated in ways that are impossible in humans, the function of the corresponding gene can be determined much more readily in a fruit fly, a nematode worm, or a yeast cell. The results are certain to provide important insights into human cancer that could not be obtained by direct research on humans. Conversely, researchers interested primarily in yeast cells will benefit from the information about yeast genes that can be derived from studies on its homologues that are initially conducted with another organism.

Even among researchers whose efforts are confined exclusively to humans, small group efforts will be encouraged. The human genome map and an ordered set of human DNA clones will be available as a resource for the use of all investigators, enabling them to concentrate on the most interesting parts of their research. In addition, new areas of research are likely to emerge as a result of this resource, particularly in relation to human health. In short, the committee believes that the mapping and sequencing project will make an important contribution to primary research conducted by small groups of independent investigators, extending their reach into currently inaccessible problems.

A project to map and sequence the human genome has many different components. In the following sections of this report, we examine implications for medicine and science (Chapter 3), mapping (Chapter 4), sequencing (Chapter 5), data handling and analysis (Chapter 6), implementation and management strategies (Chapter 7), and commercial, legal, and ethical implications (Chapter 8).

## REFERENCES

Alberts, B., D. Bray, J. Lewis, M. Raff, K. Roberts, and J. D. Watson. 1983. Molecular Biology of the Cell. Garland, New York. 1146 pp.

Alberts, B., D. Bray, J. Lewis, M. Raff, K. Roberts, and J. D. Watson. 1989. Molecular Biology of the Cell, 2nd edition, Garland, New York, in press.

Doolittle, R. F., D. F. Feng, M. S. Johnson, and M. A. McClure. 1986. Relationships of human protein sequences to those of other organisms. Cold Spring Harbor Symp. Quant. Biol. 51:447–455.

Egeland, J. A., D. S. Gerhard, D. L. Pauls, J. N. Sussex, K. K. Kidd, C. Allen, A. M. Hostetter, and D. E. Housman. 1987. Bipolar affective disorders linked to DNA markers on chromosome 11. Nature 325:783–787.

Gusella, J. F., N. S. Wexler, P. M. Conneally, S. L. Naylor, M. A. Anderson, R. E. Tanzi, P. C. Watkins, K. Ottina, M. R. Wallace, A. Y. Sakaguchi, A. B. Young, I. Shoulson, E. Bonilla, and J. B. Martin. 1983. A polymorphic DNA marker genetically linked to Huntington's disease. Nature 306:234–238.

Lewin, B. 1987. Genes, 3rd ed. John Wiley & Sons, New York. 737 pp.

St George-Hyslop, P. H., R. E. Tanzi, R. J. Polinsky, J. L. Haines, L. Nee, P. C. Watkins, R. H. Myers, R. G. Feldman, D. Pollen, D. Drachman, J. Growdon, A. Bruni, J.-F. Foncin, D. Salmon, P. Frommelt, L. Amaducci, S. Sorbi, S. Piacentini, G. D. Stewart, W. J. Hobbs, P. M. Conneally, J. F. Gusella. 1987. The genetic defect causing familial Alzheimer's disease maps on chromosome 21. Science 235:885–890.

Watson, J. D., J. Tooze, and D. T. Kurtz, 1983. Recombinant DNA: A Short Course. W. H. Freeman, San Francisco.

# 3

# Implications for Medicine and Science

## MEDICAL USES

### *A Map of the Human Genome Will Greatly Facilitate the Identification of Specific Disease Genes*

Humankind is afflicted by more than 3,000 known different inherited disorders. Taken together, these disorders affect every organ, system, and tissue in the human body. Some cause disease even before birth, whereas others are observed only in adulthood. Some are common, others rare. Although their overall impact on human health is enormous, until recently our understanding of the vast majority of these disorders has been meager. Even today we have identified the responsible gene in fewer than 3 percent of all known inherited disorders. In nearly all of these cases the disease gene codes for a known protein. For diseases in which the responsible protein has been identified, it is now regularly possible, with recombinant DNA methods, to clone the gene and begin to understand the genetic defect. In this way we have learned much about conditions such as thalassemia, sickle-cell anemia, hemophilia, Tay-Sachs disease, and familial hypercholesterolemia. However, most disorders result from mutations in genes whose protein products have not been defined. In these situations, identification of a DNA segment that is regularly altered (either by deletion, rearrangement, or point mutation) in a given disorder provides clues to identifying the disease gene. So far, the genes for three disorders—Duchenne muscular dystrophy, retinoblastoma, and chronic granulomatous disease—have been successfully identified in this manner. This approach is also making possible an ongoing search for the genes relevant to such conditions as cystic fibrosis, Huntington's disease, and familial Alzheimer's disease. These are but a small subset of the numerous Mendelian disorders for which direct genetic analysis offers the best hope of identifying the responsible genes.

The availability of various types of maps of the human genome would greatly facilitate the search for genes related to specific inherited diseases. A detailed genetic linkage map based on RFLPs would permit rapid assignment of disease loci to subchromosomal regions, perhaps at a resolution of 1 million nucleotides. The availability of DNA clone collections and a restriction map of the genome would then allow efficient comparative analysis of DNAs from normal and affected individuals to pinpoint with higher resolution the area in which the relevant gene resides. Finally, a DNA sequence of the genome would allow all putative genes in the region to be identified and would also provide a data base for evaluating sequences obtained in samples of DNA from patients. Although more complicated in its execution, similar approaches could be applied to the more common multigenic disorders, i.e., those for which more than one gene may be responsible. Examples include hypertension, some forms of cancer, diabetes, schizophrenia, mental retardation, and neural tube defects. Thus, the availability of a map and sequence would greatly accelerate the identification of disease genes and permit investigators to focus more rapidly on the nature of the gene products and their cellular roles.

### *Disease Genes Promise to Provide Important Insights into Human Biology*

An understanding of normal physiology and biochemistry has often been gained through the study of single gene disorders for which protein products have been characterized. For example, elucidation of many pathways of intermediary metabolism resulted from the examination of cells from patients in whom a single enzyme activity was abolished. Similarly, the study of individual mutant genes encoding uncharacterized products is certain to illuminate new biochemical and cellular mechanisms related to both normal human physiology and to the development of disease. The rapid identification of disease genes will enable investigators to examine in detail the protein product of such genes and their role in cellular biology. When few clues to pathophysiology exist (e.g., neurofibromatosis, polycystic kidney disease, or retinitis pigmentosa), this strategy will provide new insights into pathogenesis.

The implications of such research are likely to be extensive. In many instances, examination of an apparently rare situation may lead to a clearer understanding of normal mechanisms that may be adversely affected in other ways in more common diseases. For instance, studies of the recently isolated gene responsible for the rather rare childhood tumor known as retinoblastoma should increase our understanding of more common cancers (Dryja *et al.*, 1986; Friend *et al.*, 1986), and studies of the genes involved in an apparently uncommon type of Alzheimer's disease may explain more general features of aging (St George-Hyslop *et al.*, 1987).

### *Specific Medical Applications*

An improved capacity to identify genes related to disease will have an immediate impact on the diagnosis, treatment, and prevention of genetic disorders. As more disease genes are isolated, DNA-based diagnosis will become more common and the potential for somatic cell gene therapy will increase. Furthermore, the availability of molecular probes for specific gene loci will permit detection of the carriers of disease-associated genes. This ability will enable parents to identify the extent to which their offspring may be at risk for a genetic defect. In addition, the identification and characterization of disease genes will lead (and already has led for many genetic disorders) to improved prenatal diagnosis of serious conditions by direct DNA analysis. Finally, the ability to determine whether individuals are carriers for specific gene defects will facilitate various epidemiological investigations of the risks associated with specific environmental factors, occupational settings, or drugs.

### *Toward an Understanding of Cancer*

Cancer results from the unregulated growth of cells. What has been learned over the past decade or so, largely through the application of molecular genetic tools, is that deregulation of growth is caused by specific genetic abnormalities, i.e., mutations in growth-related genes that are either inherited or acquired during life. Inherited defects generally confer increased susceptibility to a particular form of cancer—for example, retinoblastoma, cancer of the colon, certain kidney tumors, and malignant melanoma. Only in retinoblastoma has the susceptibility gene been identified. The search for the responsible genes in other instances is in its early stages and will be greatly facilitated by detailed RFLP and DNA clone maps and the nucleotide sequence. With the susceptibility genes in hand, it will be possible to identify by testing an individual's DNA those who need special surveillance for precancerous or early cancerous changes so that appropriate treatment can be applied at an early stage of disease. It may also become possible to counter the effects of inherited susceptibility more directly once the physiological effects of the various genes are understood.

In recent years much has been learned about acquired genetic abnormalities related to cancer. During one's lifetime, the DNA in somatic cells undergoes mutation, either spontaneously or as induced by environmental mutagens. These mutations involve changes in nucleotides, rearrangements, duplications, or deletions. Some of these changes occur in genes that regulate growth. Several dozen genes are now known that, when mutated in specific ways or overexpressed, deregulate cell proliferation. Some of these abnormal genes (called oncogenes) have been found in human cancer cells and seem to contribute to their tumorigenic properties. In several instances the proteins encoded by oncogenes have been shown to be altered forms of cell growth stimulators or the cellular receptors for growth stimulators. Other oncogenes encode proteins that are involved in the intracellular response of cells to growth stimulators. As a result of these findings, primary questions regarding cell growth and human cancer have come into sharp focus: What normal human proteins are involved in cell growth and how do they act? How do changes in one or more of these proteins cause cells to grow into tumors and to spread to distant organs? What genetic mechanisms underlie these changes? What is the spectrum of oncogenes or metastasis genes present in human tumors?

The availability of a map and sequence of the human genome and of the genomes of simpler organisms will help answer these questions. It will facilitate the isolation of genes that are homologous to known growth-related genes and the identification of previously undiscovered genes that play a role in cell growth and development. The characterization of the genes and proteins that regulate cell growth and are responsible for neoplasia and metastasis of tumor cells is likely to lead to more sensitive diagnostic and prognostic tests and to new approaches to the control of cancer.

## IMPLICATIONS FOR BASIC BIOLOGY

### *What Aspects of Genome Organization Are Important for Genome Function?*

The principles of genome organization are poorly understood. The human chromosome contains functional segments that are not genes.

Specific segments are essential for the duplication of the chromosomes before cell division and for ensuring that the correct complement of chromosomes segregate into the two daughter cells. The nature of these segments within a chromosome and the mechanism by which they carry out their functions are poorly understood in mammals. A physical map of the human genome will provide the basis for experimentation into the identity and role of these and other elements.

The study of genome organization, that is, the order in which genes occur along a chromosome and their relations to various other components, will be enhanced by the existence of a physical map. For example, we do not know in most cases whether the order of genes on a given chromosome is important to their function. Is there a selective advantage to the organism to maintain the proximity of genes that are expressed together? Limited studies comparing the overall organization of genes in the chromosomes of humans and mice suggest that the organization of large blocks of genes has often been conserved, but it is not known whether this is important to their function (Sawyer and Hozier, 1986). By comparing the physical maps of a variety of organisms, it will become apparent which segments are conserved in their gene order across species and therefore are likely to have functional significance.

The detailed comparison of corresponding mouse and human DNA sequences is likely to be of special importance. Sufficient time (an estimated 70 million years) has elapsed since the divergence of mice and humans from a common mammalian ancestor for those chromosomal regions whose nucleotide sequence is not crucial for the function of the organism to differ extensively as a result of random events that change nucleotide sequences. Thus, a comparison of mouse and human sequences can reveal those regions of our chromosomes with crucial functions that are reflected as conserved (i.e., common) nucleotide sequences. Evolutionary biologists believe that changes in most of these sequences have occurred at one time or another during evolution, but because the changes were deleterious, the mutant individuals who carried such changes were eliminated from the population by natural selection. Included among the conserved sequences will be the exons of important proteins as well as the sequences in genes that regulate gene expression. Other conserved sequences whose function cannot be anticipated will no doubt be discovered in this way; their identification should eventually provide many new insights into the functions of both genes and genomes.

### *Many New Human Genes and Proteins Will Be Identified*

Only a small percentage of the human genes involved in normal development and disease have been identified to date. Mapping and sequencing the human genome will result in the identification of a large number of new genes and their encoded proteins. As one benefit, the physical map will help pinpoint the position of human genes that have been mapped to specific chromosomal locations but have not yet been isolated. Moreover, genetic studies of the mouse have revealed mutations in many genes that cause interesting pathological defects, but little is known about these genes except their location on the genetic map of mice. By knowing the specific correspondence between the physical maps of humans and mice, the corresponding gene can be identified and studied in both organisms.

There are also computer-based methods for detecting genes when the only information available is a long stretch of continuous nucleic acid sequence (Staden and McLachlan, 1982). These methods have been improving dramatically, and a human genome project will stimulate further improvement in existing computer-based tools. At present, the identification of genes and their protein products relies on several methods. First, the exons within a DNA sequence can often be predicted by identifying those segments that contain open reading frames (regions of nucleotide sequence without the "stop codons" that terminate protein synthesis) and also have codon usage biases (the preferential use of one of several codons that specifies a particular amino acid) that are consistent with other genes in that organism. Moreover, there are conserved sequences that always flank an intron.

As a second approach, genes often share homologies with one another on the basis of common evolutionary history; these homologies have been successfully exploited in a number of areas, for example, to identify related family members of lymphokines, to find new receptor proteins for neurotransmitters, and to find genes that may play important roles in pattern formation in development. Many sequence motifs that encode protein domains with a similar function have been identified, such as the common domain found in all protein kinases. These have been useful in predicting the function of unidentified gene products from their amino acid sequences. As increasing numbers of new proteins are isolated and functionally characterized, the data base available for such comparisons will be greatly increased. Many proteins contain domains that have been used over and over again in the construction of related proteins. Therefore, it should eventually be possible to discover a great deal about the structure and function of a protein from the amino acid sequence derived from its gene. Because exons coincide in many instances with protein domains, knowledge of the exon-intron structure of a gene can also provide insights into both the structure and function of the protein.

### *How Do Organisms Evolve?*

To gain a deep understanding of organisms we must understand how they evolved, and much of the evolutionary history of humans is present in our genomes. If we knew the complete DNA sequences of humans and other organisms, we should be able to trace the origins of most of our genes. However, because all mammals are constructed from similar sets of proteins, the building blocks that are used to construct a human and whale are very much the same. The many differences between mammalian species are therefore believed to depend largely on differences in the regulatory signals that control the timing, level, and cell specificity of gene expression. Thus, the orderly development of the human embryo requires that specific gene sets be activated at exactly the right place and time as new cell types arise from multipotential cells. This process is controlled at least in part by regulatory DNA sequences located near the genes. In many cases, these sequences will be homologous among those genes that are coactivated. The sequence analysis of the human genome, and its comparison with the sequence of other mammalian genomes such as the mouse, should allow us to identify very large number of regulatory DNA sequences. Moreover, one can hope to begin to understand not only the rules that govern gene regulation but also the changes that have occurred during evolution that have differentiated the human organism from our mammalian relatives.

In summary, the acquisition of the map and sequence of the human genome will expand our understanding of many basic questions in biology. To maximize this impact, it will be necessary to pursue the analysis of genomes of organisms that can be experimentally manipulated. Thus, for example, the function of regulatory sequences detected in humans can be tested by experiments in the mouse in which transgenic animals can be constructed with appropriately engineered genes. Because many crucial insights may be gained from such comparative studies, experiments in several other organisms will inevitably be required to test the function of potentially important human genes.

## REFERENCES

Dryja, T. P., J. M. Rapaport, J. M. Joyce, and R. A. Petersen. 1986. Molecular detection of deletions involving band q14 of chromosome 13 in retinoblastomas. Proc. Natl. Acad. Sci. U.S.A. 83:7391–7394.

Friend, S. H., R. R. Bernards, S. Rogelj, R. A. Weinberg, J. M. Rapaport, D. M. Albert, and T. P. Dryja. 1986. A human DNA segment with properties of the gene that predisposes to retinoblastoma and osteosarcoma. Nature 323:643–646.

Sawyer, J. R., and J. C. Hozier. 1986. High resolution of mouse chromosomes: Banding conservation between man and mouse. Science 232:1632–1635.

Staden, R., and A. D. McLachlan. 1982. Codon preference and its use in identifying protein coding regions in long DNA sequences. Nucleic Acids Res. 10:141–156.

St George-Hyslop, P. H., R. E. Tanzi, R. J. Polinsky, J. L. Haines, L. Nee, P. C. Watkins, R. H. Myers, R. G. Feldman, D. Pollen, D. Drachman, J. Growdon, A. Bruni, J.-F. Foncin, D. Salmon, P. Frommelt, L. Amaducci, S. Sorbi, S. Piacentini, G. D. Stewart, W. J. Hobbs, P. M. Conneally, and J. F. Gusella. 1987. The genetic defect causing familial Alzheimer's disease maps on chromosome 21. Science 235:885–890.

# 4

# Mapping

The genes that specify the biological heritage of each human being are arranged along chromosomes in a nearly invariant order. Consequently, simple one-dimensional maps can specify the genetic organization of the human, as well as other species. Some applications of these maps have already been described. In this chapter, the committee provides a more detailed view of the types of chromosome maps, their uses, and technical problems affecting their construction.

In considering current and future uses of maps in genetics, it is important to recognize that the exploration of the human genome is at an early stage. The roles of maps in human genetics may be expected to change with time. Over the next few years, maps will largely be used to guide the search for the DNA sequences responsible for particular genetic diseases and in genetic counseling. As systematic studies of the structure and function of the human genome expand, the role of maps in organizing information and planning new types of research will increase in importance. It would be impossible, for example, to organize systematic DNA sequencing of the human genome without precise maps of the regions to be sequenced. Even when extensive sequence data become available, maps will remain indispensable to a wide variety of genetic data, including the sequences themselves. The continued value of chromosome maps has been demonstrated for viruses whose genomes have been completely mapped and sequenced. Researchers who study such viruses keep detailed maps of the viral genome within reach at all times, but consult the sequence data less frequently. Genetic linkage maps and physical maps (even when incomplete), as well as partial sequences, have been of value in research on *Escherichia coli* (a bacterium) and *Drosophila* (a fly). In the latter, maps have been of critical importance in guiding investigators and have provided direction to the regions of interest that need to be sequenced. A similar future awaits maps of the human genome: These maps will not only be critical tools during the coming decades of discovery, but will also form a permanent part of the basic description of humankind's genetic endowment.

### *Early Cytological Mapping Efforts Depended on Examining Chromosomes Under the Light Microscope*

All types of mapping involve measuring the positions of easily observed landmarks. Until recently, the only useful physical landmarks along human chromosomes have been cytogenetic bands. When cultured human cells are treated with suitable drugs during cell division, the chromosomes are easily viewed through the light microscope as wormlike shapes. Several staining procedures developed in the late 1960s and early 1970s imprint reproducible patterns of light and dark bands on chromosomes (George, 1970). The banding pattern is believed to reflect a periodicity in the spacing of certain types of DNA sequences along chromosomes. From a mapping standpoint, this banding is important in that it allows human chromosomes to be individually recognized by light microscopy and allows an average chromosome to be subdivided into 10 to 20 regions. Banding patterns provide the basis for a physical map of the chromosomes, often referred to as a cytogenetic map. In clinical genetics, examination of the banding patterns has led to diagnosis of such conditions as the Down syndrome, a genetic disease usually caused by the presence of an extra copy of chromosome 21 (Lejeune *et al.*, 1959).

Since the late 1960s, it has been possible to assign many genes to locations on the cytogenetic map by the techniques of somatic cell genetics (Weiss and Green, 1967). In these techniques, rodent and human cells are fused to form hybrid cells that can be grown in culture. These cells generally lose all but one or a few human chromosomes, but different human chromosomes, or parts thereof, are retained in different cell lines. Chromosome banding is used to determine which portions of the human genome have been retained in particular cell lines. Consistent co-retention of a region of the genome and a human biochemical trait allows the genetic determinant of that trait to be assigned to a position on the cytogenetic map.

More than 1,000 genes and other DNA sequences have now been assigned to positions on the cytogenetic map (McKusick, 1986). Mapping activities have provided an important focus for international activities in human genetics, including studies by laboratories in at least 12 countries on 4 continents. International gene-mapping workshops have been organized every year or two since 1973. The ninth workshop was held in Paris in September 1987.

### *The Current Revolution in Genome Mapping Is Based on the Use of Recombinant-DNA Techniques*

The systematic application of recombinant-DNA technology to chromosome mapping began in approximately 1980. Since that time, it has become apparent that recombinant-DNA techniques can potentially create chromosome maps with an accuracy and level of detail that only a few years ago seemed unachievable. It is no exaggeration to say that current maps of human chromosomes compare in quality to the navigational charts that guided the explorers of the New World. Another decade of special effort directed toward mapping the human genome could yield maps comparable to the best modern maps of the earth's surface.

No single application of recombinant DNA-technology is responsible for creating this historic opportunity for progress in human genetics. Instead, the revolution in chromosome mapping has developed on several fronts, all of which are spin-offs of the extraordinary advances in DNA experimentation that took place during the 1970s. Methods of cloning DNA molecules from any organism into microbial cells, of cleaving molecules at specific sites, and of separating DNA fragments that differ only slightly in size have all contributed to present mapping capabilities. Also of major importance are DNA-probe techniques that allow a particular DNA sequence, usually obtained from a DNA clone, to be used to detect other DNA molecules with similar or identical sequences in uncloned DNA that is extracted from human or other cells. Whether chromosome mapping is being done at the level of the chromosomal DNA molecule (physical mapping) or by following the pattern in which portions of chromosomes are passed through pedigrees (genetic linkage mapping), the experimental face of chromosome mapping has changed beyond recognition since 1980.

Nevertheless, the scale of current activity is small relative to the amount of work that must be done to study the unexplored territory in the human genome. Only a major special effort directed toward systematic mapping of the human chromosomes will allow this revolution to produce, within a decade or less, a comprehensive, detailed map of the human genome.

## FUNDAMENTALS OF GENOME MAPPING

### *Physical Maps Describe Chromosomal DNA Molecules, Whereas Genetic Linkage Maps Describe Patterns of Inheritance*

Physical maps specify the distances between landmarks along a chromosome. Ideally, the distances are measured in nucleotides, so that the map provides a direct description of a chromosomal DNA molecule. The most important landmarks in physical mapping are the cleavage sites of restriction enzymes. The maps can be calibrated in nucleotides by measuring the sizes of the DNA fragments produced when a chromosomal DNA molecule is cleaved with a restriction enzyme.

Restriction mapping has not yet been extended to DNA molecules as large as human chromosomes. Physical maps of human chromosomes are now based largely on the banding patterns along chromosomes as observed in the light microscope. One can only estimate the number of nucleotides represented by a given interval on the map; furthermore, the amount of DNA present in different bands of the same size may not be constant since there are likely to be regional variations in the extent to which chromosomes condense during cell division. Nonetheless, cytogenetic maps are considered to be physical maps because they are based on measurements of actual distance.

In contrast, genetic linkage maps describe the arrangement of genes and DNA markers on the basis of the pattern of their inheritance. Genes that tend to be inherited together (i.e., linked) are close together on such maps, and those inherited independently of one another are distant. Genes from different chromosomes are inherited independently and thus are always unlinked. Genes on the same chromosome can be tightly or loosely linked or unlinked, as reflected in the probability that they will be separated from one another during sperm or egg production. The genes can be separated if the chromosome breaks and exchanges parts with the other member of the chromosome pair, a process know as crossing over or genetic exchange. The farther apart two genes are on the chromosome, the more frequently such an exchange will occur between them.

Exchange is a complex genetic process that accompanies the formation of sperm cells in the male and egg cells in the female. Unlike other cells, which contain two copies of each chromosome (except for the special case of the X and Y chromosomes in males), sperm and egg cells contain only a single copy of each chromosome. A particular sperm or egg cell, however, does not simply receive a precise copy of one of the two parental versions of each chromosome: Instead, each sperm or egg receives a unique composite of the two versions, produced by the series of cutting and splicing events that constitute genetic exchange. Indeed, the great variety of individual chromosomes that can be produced by exchange and independent assortment is responsible for much of the genetic individuality of different humans.

The order of genes on a chromosome measured by linkage maps is the same as the order in physical maps, but there is no constant scale factor that relates physical and genetic distances. This variation in scale exists because the process of exchange does not occur equally at all places along a chromosome. Nor does exchange take place at the same rate in the two sexes; hence, as maps become more accurate, there will have to be separate genetic linkage maps for males and females.

Because they describe the arrangement of genes at the most fundamental level, physical maps are gaining in importance relative to genetic linkage maps in most areas of biological research. They can never displace genetic linkage maps, however, which are distinctive in their ability to map traits that can be recognized only in whole organisms. Disease genes are particularly important illustrations of this point. Huntington's disease and cystic fibrosis, for example, have catastrophic effects on patients, but cannot be recognized in the types of cultured cells that are suitable for genetic studies. Only by studying the patterns in which these diseases are inherited in affected families has it been possible to localize the defective genes on chromosome maps. Because of the unique ability of genetic linkage mapping to define and localize disease genes, increasing the number of genetic markers available for this type of mapping should receive major emphasis in any overall program to map the human genome.

A type of physical map that provides information on the approximate location of expressed genes is a complementary DNA (cDNA) map. A gene that is expressed will produce messenger RNA (mRNA) molecules in those cells in which the gene is active (Figure 2–3). The physical mapping of expressed genes (exons) is possible by using the DNA prepared from messenger RNA in the process called reverse transcription (in which an enzyme synthesizes a complementary strand of DNA by copying an RNA molecule that serves as a template). The availability of cDNAs permits the localization of genes of unknown function, including genes that are expressed only in differentiated tissues, such as the brain, and at particular stages of development and differentiation. Because they are expressed, they are likely to be the biologically most interesting part of the genome and therefore can usefully be the focus for early sequencing. In addition, knowledge of their map locations provides a set of likely candidate genes to test once the approximate location of a gene that is altered in a particular disorder has been mapped by genetic linkage techniques.

To this point, about 4,100 expressed gene loci have been identified by all methods (McKusick, 1986). Identification of the rest of the 50,000 to 100,000 genes in the haploid genome will come eventually with complete sequencing, but can be greatly facilitated in the immediate future by the cDNA map. This map contains information of great biological and medical significance simply because it represents the expressed portion of the genome.

### *The Development of Ordered Collections of DNA Clones Is an Important Adjunct to Physical Mapping*

In theory, sensitive DNA-probe technologies make it possible to construct physical maps while cloning only a small fraction of the genome that is being mapped. In practice, however, this approach is suitable only for the coarsest level of physical mapping. At higher resolutions, most physical mapping is likely to be carried out on collections of DNA clones that have been ordered according to their positions in the original genome. The individual clones are especially useful because they provide an inexhaustible source of the DNA from each genomic region. The vectors used for DNA cloning can be plasmids, bacterial viruses, modified bacterial viruses called cosmids, or artificial yeast chromosomes. All of these types of DNA molecules are characterized by the ability to replicate exactly as autonomous units inside suitable host cells. Having ordered clone collections is also a prerequisite to most methods of sequencing the genome since the clones would provide the actual DNA fragments that would be purified and prepared for DNA sequencing.

### *Both Physical and Genetic Linkage Maps Can Be Constructed with Various Degrees of Resolution and Connectivity*

All types of mapping presuppose an inherent trade-off between the level of detail, or resolution, in a map and the extent to which the map provides a convenient overview of the mapping objective (its connectivity). An atlas of street maps for all the major cities in a state, for example, has high resolution but low connectivity. Separate maps must be presented for each city since a fully connected map of the whole state at the same resolution used for the street maps would be too big to be useful.

As a practical matter, constructing maps that combine high resolution and high connectivity is difficult. This technical challenge is likely to be the dominant problem in the systematic physical mapping of the human genome. The nature of the difficulty can be appreciated by analogy with conventional cartography. Suppose, for example, that the only two sources of data available for mapping the United States were satellite pictures of multistate regions and local property surveys. An adequate set of overlapping satellite pictures would allow construction of a fully connected, low-resolution map, whereas the local surveys would provide detailed maps of small regions. It would be extremely difficult, however, to relate the two types of data. In principle, this problem could be solved by painstakingly piecing together the local-survey maps until they covered regions large enough to discern on the satellite pictures. In practice, however, accuracy would suffer as the survey maps were pieced together, since regions such as lakes and deserts would disrupt connectivity. In general, the only powerful solution to this type of problem lies in the development of mapping methods that can achieve a series of intermediate resolutions.

In chromosome mapping, cytogenetic maps of the banding patterns seen in the light microscope correspond to the satellite pictures, whereas restriction-site maps correspond to the local surveys. Even the most extensive restriction-site maps of local regions of human chromosomes do not yet cover even a single band on the cytogenetic map. Prospects for filling in intermediate levels of the resolution hierarchy are good, but these techniques are still being developed. Ultimately, the DNA sequence will represent the physical map of the human genome at the highest possible resolution. Nonetheless, as the analogy with conventional cartography suggests, sequencing cannot stand alone: It must anchor—at the high-resolution end—a program of mapping at a whole series of resolutions.

## GENETIC LINKAGE MAPPING

### *Restriction Fragment Length Polymorphisms Are Convenient Landmarks for Genetic Linkage Mapping*

Human beings differ from one another at many points in their genomes: Some of these differences account for differences in traits such as eye color, blood type, height at maturity, or susceptibility to a particular disease. Most differences, however, have few or no consequences in terms of the appearance or function of the individual. Nonetheless, they can still be detected since they cause subtle differences in proteins or, at a minimum, in the DNA sequence. The phenomenon of multiple genetic variants at a particular site in the genome is called polymorphism. With the advent of recombinant-DNA methods and, more particularly, DNA-probe technology, a versatile type of polymorphism called restriction fragment length polymorphism (RFLP) has come to dominate human genetic linkage mapping (Botstein *et al.*, 1980; White *et al.*, 1985). RFLPs are DNA-sequence polymorphisms that result in variations in the local restriction map at particular sites in the genome. These variations are readily detected in small amounts of DNA extracted from blood samples.

The inheritance of RFLPs can be followed through families by analyzing DNA from parents and children. Because (with the exception of the X and Y chromosomes in males) each of us has two versions of each chromosome, we have two versions of each gene and DNA sequence—one inherited from each of our parents. Thus, polymorphic DNA sequences such as genes or RFLPs can be present in one person in two different forms. In such a case, the person is said to be heterozygous, carrying two different forms, called alleles, of the polymorphic gene or sequence. Heterozygosity allows investigators to track genes through families and to detect linkage. An ideal genetic marker is one that exists in so many distinct forms that every individual is heterozygous, and unrelated individuals are heterozygous for different forms. In this case, the marker can be traced unambiguously from grandparent to parent to child in every family group studied, allowing the inheritance of linked genes in the family to be traced accurately and efficiently. Actual RFLPs don't approach this ideal, but a newly discovered type of molecular marker comes much closer. These VNTRs (variable number tandom repeats) are short repeated regions that vary in length and may exist in a dozen (rather than just two) identifiable forms.

### *Genetic Linkage Mapping Requires the Study of Many People in Large Family Groups*

Two genes that are close to one another on a chromosome show tight linkage: The particular alleles of the two genes that a person inherits from one of his or her parents are almost always passed on together to that person's children. However, two genes that are farther apart but still on the same chromosome are more likely to be separated by exchange during sperm or egg production. The probability of such an exchange increases with the physical distance between the genes, thereby accounting for the observation that genes are ordered in the same way by genetic linkage and by physical mapping.

To measure the degree of exchange between two genes, the

frequency of co-inheritance of parental allele combinations must be measured on a statistically significant sample. From a practical standpoint, detection of linkage requires the measurement of the allele combinations passed from one generation to the next by at least 10 sperm or egg cells, meaning that at least five offspring must be examined from a fully informative mating, (i.e., both parents heterozygous at both sites with all parental alleles distinguishable from one another). However, an accurate measurement of the extent of linkage requires the examination of even more people. The unit of distance in genetic linkage mapping is called the centimorgan (cM), in honor of the great American geneticist Thomas Hunt Morgan. By definition, two sites that are spaced by 1 cM have a 1 percent probability of being separated by exchange during sperm or egg production. Averaged over the whole genome, 1 cM on the genetic linkage map corresponds to approximately 1 million nucleotide pairs, although the relation between genetic and physical distances varies considerably.

Great progress has been made in genetic linkage mapping with RFLPs since the concept was introduced in 1980 (Botstein *et al.*, 1980). Hundreds of RFLPs have been described, and many maps of whole chromosomes and portions of chromosomes have been published (Drayna and White, 1985). The major laboratories engaged in RFLP mapping have formed a highly effective collaboration centered around the Centre d'Etude Polymorphisme Humain (CEPH) in Paris (Marx, 1985; Dausset, 1986). In CEPH, collaborating investigators are provided with DNA from cultured cells derived from the lymphocytes of the members of 40 families having an average of approximately eight children each, as well as both parents and all four grandparents. This collection comprises approximately 600 progeny chromosome sets. By agreement among the collaborators, RFLPs that are mapped with any of the CEPH families are analyzed throughout all the families for which they are informative.

Consequently, information is steadily accumulating about the positions of recombination events in all the progeny chromosomes in the collection. The data are pooled and distributed at regular intervals to all interested investigators. This international collaboration has greatly speeded human genetic linkage mapping and lowered the entry barriers for new investigators who are interested in joining the effort. In fact, the large demand for this material makes it important to increase the number of cultured cells chosen from families that are especially useful for linkage studies.

A genetic linkage map of the entire human genome at an average resolution of about at a 10 cM was recently reported (Donis-Keller *et al.*, 1987). Current technology seems to allow construction of an

RFLP map with an average resolution of 1 cM within the next several years. This increase in resolution would require the mapping of several thousand RFLPs on a set of families larger than the current CEPH collection.

Recent innovations in human linkage mapping now allow three-point and higher multipoint mapping to be performed. This makes mapping more efficient and more like the *Drosophila* mapping that has been so productive. Maps will also be of primary importance in areas such as genetic counseling and in disease research. Obtaining markers on both sides of the genes of interest will provide more reliable information.

Genetic linkage maps of humans will require special statistical and computer techniques because humans, unlike experimental animals, often have few siblings. Computers also make it possible to do linkage analysis of complex pedigrees.

### RFLPs Are Useful for Interrelating Physical and Genetic Linkage Maps

Genetic linkage mapping allows those genes with no known cellular or molecular effects to be located on the human genome. On the other hand, physical maps describe the DNA molecules present in chromosomes. RFLP markers can easily be localized on either type of map. Not only can RFLP markers be placed on the genetic linkage map in family studies, but also, because the probes that are used to recognize RFLPs are themselves DNA molecules, their positions on a physical map can be determined in a variety of straightforward ways. Exact alignment between the genetic linkage and physical maps of the human genome at a large number of sites is therefore possible. This will greatly facilitate finding the actual DNA sequences that correspond to a gene once such a gene is localized on the genetic linkage map. In addition, making maps continuous across entire chromosomes will be easier by genetic linkage mapping, whereas maps of higher resolution (finer than a million nucleotides) will be easier to achieve by physical mapping. The more points at which the two maps can be exactly aligned, the greater the opportunity to take advantage of this complementarity, which will help solve the connectivity problem that arises when making maps of high resolution.

### A Reference RFLP Map for the Human Would Be a Critical Tool for Studying Inherited Diseases

RFLP mapping provides a powerful, comprehensive approach to the study of inherited diseases. Ideally, the centerpiece of this approach would be a reference RFLP map, at 1 cM resolution, determined from normal families. Once completed, the project of constructing such a map would provide human geneticists with a permanent archive of several thousand DNA probes that would detect polymorphisms throughout the genome at an average spacing of 1 million nucleotides. To apply this resource to the study of a particular inherited disease, an investigator would test DNA samples from families afflicted by a particular inherited disease with a uniformly spaced subset of perhaps 5 percent of these probes. Once rough linkage was tentatively detected, typically with a recombination frequency of 10 percent between the mutant gene that caused the disease and the polymorphism that was detected by the probe, the linkage could be rapidly confirmed and the position of the disease gene refined by follow-up analyses conducted with more closely spaced probes, selected to cover the region of interest thoroughly. Because the same RFLP polymorphisms are not segregating in all families, more sites are required than might seem necessary. For this reason more reference pedigrees are needed. In addition, research in highly polymorphic sites and ways of detecting them should be encouraged.

At present, genetic linkage mapping with RFLPs is often begun with essentially random probe collections; once weak linkage is detected, the refinement of the position of the disease gene is extremely laborious since new sets of probes must be developed. Nonetheless, when major resources are directed to the study of particular diseases—such as cystic fibrosis and Huntington's disease—progress can be impressive. Only a few years ago, nothing was known about the position in the genome of the gene responsible for either of these diseases, and no compelling evidence existed that either was caused by mutations in the same gene in different afflicted families. Now, as a result of the RFLP approach, both genes have been mapped with great precision and shown to have a common genetic basis in most or all cases (Gusella *et al.*, 1983; White, 1986). Equally important, the RFLP approach, because of its ability to interrelate genetic linkage and physical mapping, has laid the groundwork for locating and analyzing the actual DNA sequences responsible for the diseases by coupled strategies of physical mapping and cloning, starting with the DNA clones used to probe for the linked RFLPs.

Generalization of this strategy to the large variety of known inherited disorders could be expected to advance our understanding of basic human biology as well as to direct improvements in the diagnosis and treatment of many diseases. The reference RFLP map for the human—and its associated collection of well-tested DNA probes—would dramatically improve the efficiency of this research, allow the study of diseases in smaller family groups, and improve the practicality of studying diseases that are caused by alterations in more than one gene. The study of multigenic disorders could ultimately revolutionize medicine, since there are likely to be multigenic genetic predispositions to such common disorders as cancer, heart disease, and schizophrenia.

## MAKING PHYSICAL MAPS

### *Medium-Resolution Mapping of Restriction Sites Is Facilitated by New Methods of Preparing and Separating Large DNA Molecules*

At low resolution, cytogenetic mapping of banded chromosomes is already advanced. At high resolution, methods such as restriction-site mapping and DNA sequencing of clones are well established. Major issues of efficiency must be considered in applying these methods to the human genome, but, in principle, there are no major obstacles. However, until recently, the middle range contained a serious gap between the highest resolution achievable in cytogenetic mapping with the light microscope (10 million nucleotides) and the lowest resolution achievable by restriction-site mapping (10,000 nucleotides).

At present, prospects of bridging this 1,000-fold gap in resolution to connect the two types of maps by increasing the resolution of cytogenetic mapping are limited. Until recently, two substantial obstacles existed to bridging it from the other direction by extending restriction-site mapping to lower resolutions (and thus longer distances). The first obstacle was a lack of restriction enzymes that cleave human DNA infrequently enough to produce the very large DNA fragments needed for low-resolution mapping. The second was an inability to separate and measure the sizes of DNA fragments appreciably larger than 20,000 nucleotides. During the past 5 years, major progress has been made toward solving both of these problems. Restriction enzymes have been discovered that cleave DNA into fragments with average sizes ranging from 100,000 to 1 million nucleotides. In addition, a method known as pulsed-field gel electrophoresis, which allows the separation of DNA fragments as large as 10 million nucleotides, has been introduced (Schwartz and Cantor, 1984).

Now that it is possible to generate, separate, and measure large DNA fragments, a variety of ways of constructing restriction-site cleavage maps exist. Cleaving a DNA genome infrequently at specific sites with appropriate restriction enzymes produces many large DNA fragments of different sizes. These fragments can then be separated from each other by electrophoresis through agarose gels. The DNA bands that result can be seen either by direct DNA staining or by nucleic-acid hybridization with appropriate DNA probes. (The latter technique takes advantage of the specificity of complementary base-pairing between two DNA strands, which allows one highly radioactive DNA molecule—the DNA probe—to be used to find its one complementary partner in a mixture that contains millions of other DNA molecules.) Although these methods allow different fragments of chromosomes to be separated and their contents of probe sequences to be determined, they provide no information regarding the order of these fragments along the chromosome. However, the 50 to 500 different large fragments produced from each human chromosome can be ordered by an extension of such analyses. One way involves cutting the genome at two distinct sets of sites with two different restriction enzymes, a procedure that generates two families of large DNA fragments that overlap. The fragments that are neighbors in the genome can then be identified with appropriate DNA probes since two overlapping fragments will hybridize to the same probe. In another method, only a single restriction enzyme is used to produce the large DNA fragments. In addition, however, a set of small DNA probes, called linking probes, is generated by selectively cloning the short segments of DNA that surround each of the cleavage sites for the restriction enzyme used to make the large fragments. Because linking probes contain sequences from both sides of a particular restriction site, each should hybridize to two different large fragments when used as a DNA probe, thereby demonstrating that these particular large fragments are neighbors in the genome (Poustka and Lehrach, 1986).

The largest DNA molecule that has been mapped with restriction enzymes that cleave DNA infrequently is the single chromosome of *E. coli* (4.7 million nucleotides) (Kohara *et al.*, 1987; Smith *et al.*, 1987). The average spacing of the mapped sites is approximately 200,000 nucleotides. Progress in achieving an *E. coli* map at higher resolution, largely by analyzing ordered sets of DNA clones, is also proceeding rapidly.

The smallest human chromosome is 10 times as large as the *E. coli* chromosome. Although difficult to construct, its physical map could be determined by methods that are generally similar to those applied to *E. coli*. In principle, such an effort would best be carried out after the human chromosomes were separated from each other, to prevent the DNA fragments of the other chromosomes from complicating the analysis of the one chromosome of interest. In recent years, progress in chromosome-separation technology has been impressive, but expert opinion remains divided as to whether the final samples are pure enough and contain enough DNA to have a major impact on physical mapping projects. The chromosomes to be separated are isolated from human cells undergoing division, a stage of the cell's life cycle when the chromosomes are condensed and stable. They can be separated according to size by flow cytometry—a method in which the amount of DNA present in condensed chromosomes is analyzed while the chromosomes flow one by one through a small tube. Computer-controlled systems allow each individual chromosome to be diverted to a designated collection tube depending on its DNA content. DNA samples prepared from chromosomes separated in this way have already served as an important source for producing clone collections that are highly enriched for the DNA sequences of a particular human chromosome.

### *High-Resolution Mapping of Restriction Sites Will Require the Use of Ordered Collections of DNA Clones*

The purification of human chromosomes can only moderately decrease the complexity of the DNA samples used for mapping. In contrast, cloning techniques offer large decreases in complexity: Through chromosome separation, the complexity of the samples can be reduced 10- to 100-fold, whereas cosmid cloning reduces the complexity of individual samples 100,000-fold. Furthermore, unlike separated samples of human chromosomes, DNA clones will replicate in microbial hosts, thereby allowing the production of as much DNA as needed. For these reasons, doing as much physical mapping as possible on cloned DNA has overwhelming advantages. Particularly for high-resolution mapping, the preferred source of DNA samples for physical mapping will be ordered collections of DNA clones—a set of cloned DNA fragments that have been sufficiently analyzed that they can be arranged to reflect the order of their corresponding DNA fragments on the original chromosomes. Since the clones are usually generated in a way that produces cloned DNA fragments that start and stop at random sites along the chromosome, each member of the collection will normally overlap extensively with several neighbors, and the entire collection will have considerable redundancy (i.e., any segment of the chromosome will be represented in several different clones).

### *Fingerprinting Methods Can Be Used to Order DNA Clones*

Preparing an ordered-clone collection involves cloning DNA fragments as molecules that can replicate in a microbial host, determining the order of these fragments in the genome, and propagating the fragments in pure form to make them widely available for subsequent analysis. Much can already be done in these respects, and the prospects for rapid advancement of technical capabilities are good. The properties of the cloned DNA fragments can then be used to reconstruct their original order in the genome. For a set of random clones, some clones will partially overlap the region of the genome covered by other clones. A characteristic of the overlapping region can be measured, such as the detailed pattern of cutting by a set of restriction enzymes. This analysis is performed for a large number of clones individually, and then a computer search of the patterns is used to place clones in order (neighboring clones are those that share part of their patterns). This method is called fingerprinting, since the identifying DNA characteristics of each cloned segment are analogous to a fingerprint of the DNA fragment.

Fingerprinting methods have recently been used successfully to order large numbers of cloned DNA segments in yeast, *E. coli*, and nematode genomes (Coulson *et al.*, 1986; Olson *et al.*, 1986; Daniels and Blattner, 1987; Kohara *et al.*, 1987). In principle, this method should provide an efficient way to group DNA clones into contiguous regions that cover 90 percent or more of the genome. A common problem, however, is that the matching of contiguous segments proceeds rapidly at first and then slows. Finishing the process by using DNA-probe techniques to find the clones needed to fill in the map then becomes time-consuming and tedious. The unexpectedly large number of gaps have two principal explanations: (1) Not all overlapping segments are being recovered because of biases inherent in the DNA cloning procedures used, and (2) the fingerprint information collected for the overlapping DNA segments lacks sufficient precision to distinguish all DNA fragments from each other unambiguously. Progress in both areas may be expected as a wider variety of cloning systems are explored and more sophisticated fingerprinting methods are developed. For example, alternatives to the use of restriction enzyme cutting patterns as the fingerprint are being explored (Poustka *et al.*, 1986).

### *The Optimal Method for Preparing Ordered Collections of DNA Clones Is Not Yet Clear*

Although the general principles of working with ordered collections of DNA clones are well established, the technology is in a state of flux. A promising recent development is the demonstration that yeast can be used as host cells for cloning large human DNA segments.

Several laboratories have shown that DNA fragments as long as 500,000 nucleotides can be cloned as artificial chromosomes in yeast. These fragments are 10 times the size of the fragments that can be cloned with current bacterial-host systems (Burke *et al.*, 1987). Further development of systems for cloning large DNA molecules will greatly enhance the efficiency of ordering DNA fragments. For example, it should be possible to prepare DNA clone collections by using a single restriction enzyme that cuts DNA infrequently; this procedure would generate a single family of large DNA fragments that are then cloned. This family would be much less complex than the collection of randomly cut clones required for the fingerprinting method. A second set of short DNA clones that specifically includes all the rare restriction sites that were cut to make the large fragments could then be used as linking probes to establish the continuity between adjacent large fragments, thereby allowing the large fragments to be ordered along the genome.

The cDNA clones representing the transcribed regions of the genome represent an alternative source of probes that could be used to demonstrate the adjacency of large cloned fragments. Because cDNA clones are made by reverse transcription of mRNAs, they lack the intron sequences that interrupt the exons in the genomic DNA. The exons that have been joined together in the cDNAs will often be encoded by the DNA from more than one large genomic fragment, so that DNA probes prepared from cDNAs can be used to order the fragments from adjacent portions of the genome. This method has the advantage that the cDNA clones are themselves of special interest since they represent the portion of the genome that is selectively expressed in cells.

Still another source of useful probes would be a set of RFLP DNA probes that have been ordered by genetic linkage analyses of standard families. An RFLP map with a 1-cM resolution would provide markers separated by 1 million nucleotides, on average. If a DNA clone collection of human genome fragments that averaged several million nucleotides in size could be constructed, it could be readily ordered with these markers.

For certain methods at least, the task of ordering the DNA clones obtained from the human genome is complicated by the considerable repetition of DNA sequences in the genomes of higher organisms. These sequences are largely absent from the *E. coli*, nematode, and yeast genomes from which ordered clone collections have thus far been prepared. Additional problems are expected from the instability of selected clones observed when *E. coli* serves as the host for cloned DNA; it is too early to know whether these problems will also apply to the newer yeast cloning systems. For all these reasons, it is uncertain which cloning and linking methods will prove most effective for a human genome project. Further methodological developments could even supplant all present methods.

## IMMEDIATE APPLICATIONS OF CHROMOSOME MAPS

A number of important applications of chromosome maps could be pursued even while the various mapping activities are progressing. We have already discussed how even a partial map can be expected to facilitate the isolation of specific human disease genes. Maps will also support early sequencing efforts. The lower resolution physical maps will provide a framework within which to organize the highly fragmentary sequence data that will be generated by these initial sequencing efforts, while the ordered-clone collections will provide the actual fragments that are subcloned for final sequencing (see Chapter 5).

Chromosome maps can also be usefully applied to begin a systematic assignment of expressed genes to map positions. Most DNA in the human genome is either not part of an expressed gene or is in one of the many intervening sequences (introns) that separate the protein-coding portions of expressed genes. As previously stated, the cloning of cDNA produces only the coding DNA sequences present in expressed genes (the exons and not the introns). It is possible to make large collections of cDNA clones derived from the genes that are expressed in particular tissues or at a particular stage of development and differentiation and to embark on the systematic assignment of each expressed gene to a map position on the chromosomes. Methods are being developed to avoid the standard problem with cDNA, which is that genes expressed at a low level are often missed, whereas genes expressed at a high level produce much mRNA and therefore are obtained repeatedly as cDNA clones. These methods aim at producing "normalized" cDNA libraries, in which each expressed DNA sequence is equally represented.

Initially, the map assignments for the expressed genes could be based on the existing cytogenetic map and could be carried out by somatic cell genetic techniques, as well as by in situ hybridization of cDNAs to chromosomes. As the physical mapping and sequencing of the genome proceeded, it would require relatively little effort to refine these map assignments.

## CONCLUSIONS AND RECOMMENDATIONS

Methods for physical and genetic linkage mapping have developed steadily and impressively over the past three decades. Today, low-resolution genetic linkage maps and cytogenetic maps exist for much of the human genome. During the past few years, these maps have led to the identification of genes or chromosome segments involved in several human diseases. These advances underscore the extent of past progress in genome mapping and the promise that it holds for contributing to improved human health.

### *Recent Breakthroughs Have Set the Stage for Large-Scale Mapping*

Breakthroughs in mapping methods during the past several years have made it possible to construct chromosome maps of unprecedented completeness, accuracy, and detail. These breakthroughs include the development of techniques that have allowed 100-fold larger DNA molecules to be separated and manipulated than previously possible. In addition, new and powerful methods for following the inheritance of arbitrary segments of chromosomes through human pedigrees are available. Both physical and genetic linkage mapping have been invigorated by these developments, and important synergism has arisen between these two approaches to genomic mapping. Consequently, the goal of developing complete physical and genetic linkage maps of the human genome in a relatively short time is now realistic. These maps would be useful in their own right and would pave the way toward constructing the ultimate human map—the complete DNA sequence of the human genome.

The task of making a human genome map will by no means be easy. The longest complete physical map that has been constructed to date is for the *E. coli* chromosome. This map is only 1/640 the size of the human genome. The *E. coli* mapping benefited from an enormous base of knowledge on the bacterium accumulated during 40 years of intensive study. For example, approximately 1,000 genes have been assigned to positions around the *E. coli* chromosome, whereas a comparable region of the human genome, on average, contains a single known gene. Even after the genetic linkage mapping is completed at the 1-million nucleotide resolution recommended in this report, an *E. coli*-sized region of the human genome would contain only a handful of genetic markers. Thus, constructing a physical map of even the smallest human chromosome with today's technology would require a substantial effort.

It is anticipated that the most difficult aspect of the physical mapping will be the achievement of long-range connectivity. Although it is likely that a large proportion of the human genome could be mapped at a resolution of a few thousand nucleotides simply by relying on the fingerprinting of overlapping DNA clones, so many gaps would likely be left that the connectivity achievable by this approach would be poor. The committee believes that the utility of the physical map will increase dramatically as its connectivity improves. Consequently, attaining high connectivity in the physical map should be a major priority of the overall human genome project.

Because the technology needed for genetic linkage mapping with RFLPs is more advanced than that for physical mapping, an immediate emphasis should be placed on completing the genetic linkage map. A project with the goal of attaining of a fully connected map with an average resolution of 1 cM is strongly recommended. This goal would require that a few thousand new RFLPs be identified and mapped by classic linkage analysis on DNA samples from a set of three-generation families. Such an effort, which could begin immediately, would be expected to require several years to complete and to cost approximately $40 million.

### *Different Mapping Methods Should Proceed in Parallel*

A critical feature in all mapping is that the results from different methods are additive and corroborative. For example, the restriction-site maps, the cDNA maps, and ordered DNA clone collections go hand in hand since each helps construct the other. The use of one of these maps to study human disease also requires a genetic linkage map. In turn, efforts to construct linkage maps at higher resolutions will be assisted by the existence of corresponding physical maps. Thus, no single strategy is best overall. All types of mapping need to be coordinated as part of a human genome project.

The natural tendency of researchers to press forward with the detailed analysis of chromosomal regions of particular interest should be encouraged. The committee specifically recommends against a centrally imposed plan to proceed from lower to higher resolution as is implicit, for example, in proposals to complete the entire physical map before initiating pilot sequencing projects. Such sequencing projects will no doubt begin with the sequencing of large chromosomal regions of particular biological interest.

### *The Improvement of Physical Mapping Techniques Should Be Closely Coupled to Actual Attempts to Map Large Genomes*

Experience teaches that the practical problems facing large-scale mapping efforts become clear only when attempts are made to apply new methods to actual map production. Many approaches that seem ideal in theory fail for reasons that cannot be foreseen. In addition, the day-to-day press of practical problems drives the development of useful new technology. Thus, the committee recommends that actual mapping efforts be supported now on a substantial scale.

Nonetheless, a major initial focus of most laboratories involved in physical mapping projects is likely to be the development of techniques. Despite recent advances, many limitations on physical mapping methods still exist. For example, DNA fragments as large as 10 million nucleotides can be handled, but only with considerable difficulty, and such large fragments cannot yet be cloned. Ordered DNA clone collections have been started, but not completed, for several organisms with genomes that are at most 1/50 the size of the human genome. Advanced technology, such as handling larger DNA pieces, can expedite the preparation of such clone collections. In addition, the stability of the cloned DNA fragments is a major concern, since once the effort is devoted to constructing an ordered DNA clone collection, one should be able to count on it as a permanent resource for future studies.

### *Specific Improvements That Will Facilitate Map Construction and Usefulness Can Be Identified*

In each aspect of mapping, major improvements in technology seem likely to emerge over the next few years. These improvements, which should be major initial goals of the human genome project, will include increased DNA size range, increased resolution, diminished cost, and improved accuracy. Some of the specific target areas include improving or creating methods for:

- Physically separating intact human chromosomes.
- Isolating and immortalizing identified fragments of human chromosomes in cultured cell lines.
- Cloning complementary DNA from low-abundance messenger RNA and obtaining "normalized" cDNA libraries.
- Cloning large DNA fragments.
- Purifying large DNA fragments.
- Separating large DNA fragments with higher resolution.
- Ordering the adjacent DNA fragments in a DNA clone bank, including mathematical and statistical work that would aid in map construction.
- Automating various steps in DNA mapping, including DNA purification and hybridization analysis, and handling of many different DNA samples simultaneously.
- Data recording, storage, and analysis, with attention to the mathematical and statistical problems of optimizing physical mapping

and sequence assembly and to the application of statistical methods of database quality control.

In addition, expanded collections of CEPH-like, three-generation families from which DNA could be distributed for genetic linkage studies will be important in facilitating map construction.

Because the technology is still in its infancy, support should be directed to those research groups judged to have the greatest ability to develop technology, rather than to routine production centers staffed mainly by technicians.

## REFERENCES

Botstein, D., R. L. White, M. Skolnick, R. W. Davis. 1980. Construction of a genetic linkage map in man using restriction fragment length polymorphisms. Am J. Hum. Genet. 32:314–331.

Burke, D. T., G. F. Carle, and M. V. Olson. 1987. Cloning of large segments of exogenous DNA into yeast by means of artificial chromosome vectors. Science 236:806–812.

Coulson, A., J. Sulston, S. Brenner, and J. Karn. 1986. Toward a physical map of the genome of the nematode *Caenorhabditis elegans*. Proc. Natl. Acad. Sci. U.S.A. 83:7821–7825.

Daniels, D. L., and F. R. Blattner. 1987. Mapping using gene encyclopaedias. Nature 325:831–832.

Dausset, J. 1986. Le centre d'etude du polymorphisme humain. Presse Med. 15:1801–1802.

Donis-Keller, H., P. Green, C. Helms, S. Cartinhour, B. Welffenbach, K. Stephens, T. P. Keith, D. W. Bowden, D. R. Smith, E. S. Lander, D. Botstein, G. Akots, K. S. Rediker. T. Gravius, V. A. Brown, M. B. Rising, C. Parker, J. A. Powers, D. E. Watt, E. A. Kauffman, A. Bricker, P. Phipps, H. Muller-Kahle, T. R. Fulton, S. Ng, J. W. Schumm, J. C. Braman, R. G. Knowlton, D. F. Barker, S. M. Crooks, S. E. Lincoln, M. J. Daly, and J. Abrahamson. 1987. A genetic linkage map of the human genome. Cell 51:319–337.

Drayna, D., and R. White. 1985. The genetic linkage map of the human X chromosome. Science 230:753–758.

George, K. P. 1970. Cytochemical differentiation along human chromosomes. Nature 226:80–81.

Gusella, J. F., N. S. Wexler, P. M. Conneally, S. L. Naylor, M. A. Anderson, E. R. Tanzi, P. C. Watkins, K. Ottina, M. R. Wallace, A. Y. Sakaguchi, A. B. Young, I. Shoulson, E. Bonilla, and J. B. Martin. 1983. A polymorphic DNA marker genetically linked to Hungtington's disease. Nature 306:234–235.

Kohara, Y., K. Akiyama, and K. Isono. 1987. The physical map of the whole *Escherichia coli* chromosome. Cell 50:495–508.

Lejeune, J., M. Gauthier, and R. Turpin. 1959. Les chromosomes humains en culture de tissues. C. R. Hebd. Seances Acad. Sci. 248:602–603.

Marx, J. L. 1985. Putting the human genome on the map. Science 239:150–151.

McKusick, V. A. 1986. Mendelian Inheritance in Man: Catalogs of Autosomal Dominant, Autosomal Recessive, and X-Linked Phenotypes, 7th ed. Johns Hopkins University Press, Baltimore.

Olson, M. V., J. E. Dutchik, M. Y. Graham, G. M. Brodeur, C. Helms, M. Frank, M. MacCollin, R. Scheinman, and T. Frank. 1986. Random-clone strategy for genomic restriction mapping in yeast. Proc. Natl. Acad. Sci. U.S.A. 83:7826–7830.

Poustka A., and H. Lehrach. 1986. Jumping libraries and linking libraries: The next generation of molecular tools in mammalian genetics. Trends Genet. 2:174–179.

Poustka, A., T. Pohl, D. P. Barlow, G. Zehetner, A. Craig, F. Michaels, E. Ehrich, A.-M. Frischauf, and H. Lehrach. 1986. Molecular approaches to mammalian genetics. Cold Spring Harbor Symp. Quant. Biol. 51:131–139.

Schwartz, D. C., and C. R. Cantor. 1984. Separation of yeast chromosome-sized DNAs by pulsed field gradient gel electrophoresis. Cell 37:67–75.

Smith, C. L., J. F. Econome, A. Schutt, S. Klco, and C. R. Cantor. 1987. A physical map of the *Escherichia coli* K12 genome. Science 236: 1448–1453.

Weiss, M. C., and H. Green. 1967. Human-mouse hybrid cell lines containing partial complements of human chromosomes and functioning human genes. Proc. Natl. Acad. Sci. U.S.A. 58:1104–1111.

White, R. 1986. The search for the cystic fibrosis gene. Science 234:1054–1055.

White, R., M. Leppert, D. T. Bishop, D. Barker, J. Berkowitz, C. Brown, P. Callahan, T. Holm, and L. Jerominski. 1985. Construction of linkage maps with DNA markers for human chromosomes. Nature 313:101–105.

# 5

# Sequencing

The nucleotide sequence of a genome is its physical map at the highest level of resolution. It provides all the information that goes into making up an individual's genetic complement, and no two individuals (except identical twins) share the same genome sequence. Rather, every human contains a duplicate copy of every chromosome, with about a 1.0 percent difference between the sequence of each of his or her two homologous chromosomes (that is, the average person in a population is heterozygous for about 1.0 percent of the nucleotide pairs, or approximately 30 million pairs). These differences have arisen by mutations accumulated over the course of evolutionary time, and most of them do not affect the normal functions of the individual. Any sequence derived from the human genome will be a prototype—a blueprint that will lay out the basic organization and sequence of the genes on the chromosomes. This prototype may be derived by forming a composite of regional sequences from many individuals; it need not represent the complete sequence of any one person. The nature of individual variation will become apparent when regions of interest are compared among individuals.

In 1971, the first nucleotide sequence was obtained directly from DNA with the determination of the 12-nucleotide-long cohesive ends of the bacteriophage λ (Wu and Taylor, 1971). Since that time, with the advent of rapid techniques, about 15 million nucleotides of DNA sequence have accumulated in the GenBank database (see Chapter 6), of which over 2 million are from human DNA (Howard Bilofsky of Bolt Beranek and Newman, Inc., personal communication, 1987). This figure represents approximately 0.07 percent of the human genome. Thus, although human genomic sequencing has already begun, unless a special effort is initiated, the entire sequence will not be available for many decades, if ever.

## WHY SEQUENCE THE ENTIRE HUMAN GENOME?

There is general agreement in the biological sciences community that a physical map of the human genome, as represented in a set of cloned overlapping DNA fragments, is a worthwhile goal. There is much less consensus on the advisability of embarking on the determination of major amounts of its nucleotide sequence. Three kinds of reservations are often voiced:

- Since the amount of useful protein-coding information in the genome is estimated to be 5 percent or less, a great deal of effort would be expended in determining the order of nucleotides of no apparent significance. If massive amounts of sequencing are to be done, why not sequence only large libraries of cDNAs instead?
- Even if only cDNAs were sequenced, we would lack the ability to utilize the vast amount of sequence information generated. The problem will be even worse with a complete genome sequence. Therefore, the limited amount of usable knowledge gained is not worth the anticipated cost.
- Even if the project is worthwhile, the intensive effort required will divert funds from other research aimed at understanding the structure and function of genes in all organisms, and, therefore, there will be a net loss rather than a net gain of important biological information.

To address the first point, one must consider whether it would be less difficult to identify, before sequencing, the 5 percent of the genome that actually encodes proteins than to sequence the entire genome. Given the present state of sequencing technology, this is certainly the case; for this reason, we would anticipate that most human genome sequencing in the immediate future will be carried out on cDNA clones, which represent the expressed DNA sequence. However, it seems fair to assume that by the time sequencing begins on a massive scale, the technology will have matured so far that inserting a preliminary step that discriminates among genes, intergenic regions, and introns—which will presumably involve sorting out all the repeated isolates of the same DNA clones—will be less efficient than sequencing large regions from ordered genomic DNA clone libraries without reference to their contents. This, of course, assumes major technological advances in sequencing, as will be described subsequently.

Another reason to sequence more than just cDNAs is that sequencing the entire genome is certain to reveal unsuspected sequences having important functions. For example, one of the great challenges of a genome sequencing project is to identify potentially important functional domains involved in gene regulation and chromosome organization. The identities of such sequences will be elicited by multiple analytical approaches and will require sequence comparisons between the analogous intergenic regions in multiple species (including human versus mouse) and the recognition of unusual patterns of sequence within a single organism. As one example, the comparative sequencing of 3,500 nucleotides of a regulatory region from the *engrailed* gene of two different *Drosophila* species has revealed the presence of more than 50 short blocks of evolutionarily conserved sequences, most of which are suspected to represent the binding sites for different gene regulatory proteins (J. Kassis and P. O'Farrell, University of California, San Francisco, personal communication, 1987). Determining the function of each of the sequences will require experimental testing based on the sequence analysis, which can pinpoint even short sequences that deserve serious investigation by virtue of their conservation during evolution.

Sequence comparisons among different species will pick up genes readily as evolutionarily conserved sequences in the genomes. However, such comparisons are rarely necessary for picking out coding sequences since existing analytical tools are adequate to identify them within a single DNA sequence. The standard procedure is to use computer programs to identify open reading frames, which are regions of nucleotide sequence lacking the stop codons that terminate a protein sequence. Practical experience shows that this information, when combined with codon usage patterns and other characteristics, allows one to identify virtually all genes in a nucleotide sequence, even though short exons will occasionally be missed (Staden and McLachlan, 1982). Accurate specification of the coding sequences can then be obtained by a standard experimental analysis of the corresponding clones in a collection of cDNAs. Some of the genes discovered will have immediate significance to the biomedical community because they are associated with a disease. Many others will be analyzed and found to contain homologies to existing gene families, an immediate clue to their possible function. As more gene sequences are determined, such relations among genes will be found with increasing frequency (as has recently happened, for example, among genes that encode cell surface proteins that bind specific protein molecules that are involved in cell signaling), and entirely new gene families will be identified as well (Doolittle *et al.*, 1986).

Critics will rightly point out that a complete human genome sequence will make such a huge number of genes—perhaps as many as 100,000—directly accessible that the function of the vast majority of them will remain unknown for many decades after the genome has been completely sequenced. Why then should one devote extra resources to speeding up the completion of the sequencing effort? The committee feels that much is to be gained from having a complete catalog of human gene sequences that does not require knowing the function of most of the individual genes. For example, scientists interested in the signaling actions of cyclic nucleotides will immediately be able to recognize a large group of genes that are likely to produce proteins that bind cyclic nucleotides. Specific antibodies can be prepared to each of these proteins and used to test for the role of each of them in any signaling pathway of interest. Whole families of proteins that are likely to mediate the signaling effect of calcium ions can be identified in a similar way. Likewise, a large group of candidate human genes will be immediately available as potential analogs of any newly discovered yeast, nematode, or *Drosophila* protein, for example. Other novel uses of the genome sequence data, unforeseen at present, will be developed by individual scientists, just as many of the most important current uses of recombinant-DNA technology were not foreseen by its early developers. In short, we anticipate that the genome sequence will serve as a basic "dictionary" that catalyzes striking advances in our understanding of cells and organisms.

In response to the third criticism, the committee specifically recommends that the sequence of the human genome be determined in parallel with analogous sequencing of the genomes of the other organisms needed to interpret the human data. Thus, the basic research on these model organisms should be closely integrated with data on humans. In addition, the project must have independent funding so that it does not divert funds from ongoing basic studies, particularly those trying to understand the function of genes in in all organisms, because it is ultimately such research that will make the information on the human genome interpretable.

Finally, a concerted sequencing effort will benefit a wide range of biological investigations. By pushing the development of sequencing technology and establishing sequencing centers, inexpensive sequencing will become available to anyone who has a legitimate need for it. In this way the envisioned project will free individual laboratories from the routine and currently labor-intensive effort of sequencing their few genes of interest—a necessary prelude to studies of gene function and regulated expression. It is extremely inefficient for each laboratory to set up the facilities needed to sequence the 100,000 to 1 million nucleotides that it finds of interest. Rather, the recommended project grows out of the recognition that elucidating nucleotide sequences (as distinct from sequence analysis) is ideally an exercise of production, not of research.

Accumulating large amounts of DNA sequence data will have an impact on the biological community in other ways as well. The information contained within the genome sequence will allow full investigation into the nature and extent of polymorphism, or diversity (see Chapter 4), in the genes in the human population. Once genes with widespread diversity (such as the major histocompatibility antigens and T-cell receptor genes) are identified, comparative sequencing of a single gene or gene family in many individuals will naturally follow. Finally, the availability of structural information on a variety of genes will stimulate efforts to correlate protein coding domains, or exons, with protein folding domains. It has been proposed that the segments of proteins encoded by individual exons arose during evolution as small protein units capable of independent folding and that they have assembled into multifunctional proteins as independent domains (Gilbert, 1978, 1985). By studying these correlations, one may learn much about the rules that govern the secondary and tertiary structure of proteins. Such spin-offs will be of great value to the biological community and are meant to augment its activities—not to detract from them.

## CURRENT TECHNOLOGY IN DNA SEQUENCING: CHEMICAL AND ENZYMATIC METHODS

Any project to sequence a large genome with many repeated sequences would not start with short, randomly selected genome fragments, even though this is the easiest way to obtain a large amount of sequence information quickly. Most of the sequences obtained in this way would be short (perhaps 200 to 600 nucleotides), and millions of gaps would be left between them. Because most genes in humans extend for many thousands of nucleotides (Table 2–1), little information of biological value can be obtained from a collection of such short sequences. For this reason, sequencing would normally begin with a large cloned segment of DNA that would be sequenced completely. Such a DNA segment must first be subcloned into smaller, more manageable fragments. This can be done by one of three methods:

- Generate a detailed restriction map, and determine from the map the identity of each subclone and its relation to the whole.

- Beginning at one end of the large segment, generate a series of successively smaller DNA fragments by a limited removal of nucleotides from the end with exonucleases (enzymes that hydrolyze the phosphodiester bonds that join nucleotides together starting at a chain end); clone the remaining DNA to produce a series of clones of known origin.
- Generate a totally random series of overlapping subclones, whose relationship to one another is revealed only after their sequencing.

Large sequencing projects often mix all three strategies. One sometimes begins by randomly sequencing fragments and follows with directed sequencing of specific subclones as the gaps are located. All sequencing strategies require some redundancy in the form of overlapping sequences in order to merge the results of several determinations from different subclones and to provide a check on the accuracy of the sequence, which requires the sequencing of both DNA strands as a cross-check on systematic errors. The subcloning method will determine to a large degree the amount of redundancy in the data. Although time-consuming during the subcloning process, the first and second subcloning methods ultimately require that any single segment be sequenced only about three times. The third method, because one is sequencing subclones at random, generally requires that each segment be sequenced approximately 10 times; however, methods are available to specifically select missing clones, after a three-fold coverage, which reduce the amount of redundant sequencing (Sanger *et al.*, 1982).

The ability to sequence large stretches of DNA became a reality in the middle to late 1970s with the independent development of two techniques. One of these, developed by Sanger and his colleagues at the Medical Research Council in Cambridge, England, is a method called enzymatic sequencing (Sanger *et al.*, 1977). The unknown sequence is subcloned into a single-stranded DNA virus, and DNA synthesis is initiated from a primer sequence adjacent to the unknown sequence. This method utilizes the principle that when appropriately designed chain-terminating analogs of the four DNA nucleotides (A, G, C, and T) are incorporated into DNA by DNA polymerase, synthesis of the growing DNA chain is terminated. For example, if the synthesis of DNA molecules begins at a fixed point on a template in the presence of a low concentration of the A analog, the analog will infrequently be incorporated instead of the normal A nucleotide at any one position. However, when incorporation occurs, the synthesis of the chain stops. Thus a nested set of DNA fragments that terminate at every A nucleotide in the unknown sequence is generated.

By correlating the length of the terminated chains with the identity of the base analog that was present in the reaction, one can determine the order of the nested DNA fragments and, hence, the corresponding nucleotide sequences (Figure 5-1). At present, this method dominates DNA sequencing applications primarily because once the subclones are generated the procedure involves only a few simple steps.

The second technique, which is referred to as chemical sequencing, was developed by Maxam and Gilbert at Harvard University (Maxam and Gilbert, 1977). It uses chemicals that break the DNA chain at specific nucleotides. The DNA molecule is labeled at one end with a radioactive tag. It is then cleaved with each chemical separately in such a way as to generate breaks infrequently at any given nucleotide. As in the enzymatic sequencing technique, the DNA fragments are separated according to size, and the sizes are correlated with the nucleotide that is cleaved (Figure 5-2). This method is generally more time-consuming than the enzymatic sequencing method, but it often produces fewer ambiguities in the interpretation of the data.

Both methods generate mixtures of specific DNA fragments that are separated by polyacrylamide gel electrophoresis—a technique that can resolve fragments that differ in size by a single nucleotide. When radioactively labeled DNA fragments are used, they are detected by exposing the gel to an x-ray film. That film, which has imprinted upon it a ladder of bands distributed over four parallel lanes representing the four nucleotides of DNA, must be interpreted or read by an experienced person and the data must be entered into a computer. Machines have been developed to expedite this process through the use of a stylus attached to a computer that points to each band on the x-ray film. The computer then registers the position and translates it into one of the four nucleotides of DNA. Attempts are now under way to develop x-ray film scanners capable of reading such films directly. In addition, automatic methods that use fluorescent labels have been introduced. It is critical that other strategies for reducing the human labor and error involved in this process be developed if the human genome is to be sequenced in a timely manner.

## THE DIFFICULTY OF DETERMINING THE SEQUENCE OF THE HUMAN GENOME WITH CURRENT TECHNOLOGY

What constrains efforts to embark immediately on a large-scale human genome sequencing project? The cost and inefficiency of current DNA sequencing technologies are too great to make it feasible to contemplate determining the 3 billion nucleotides of the DNA sequence in the human genome within a reasonable time. The largest

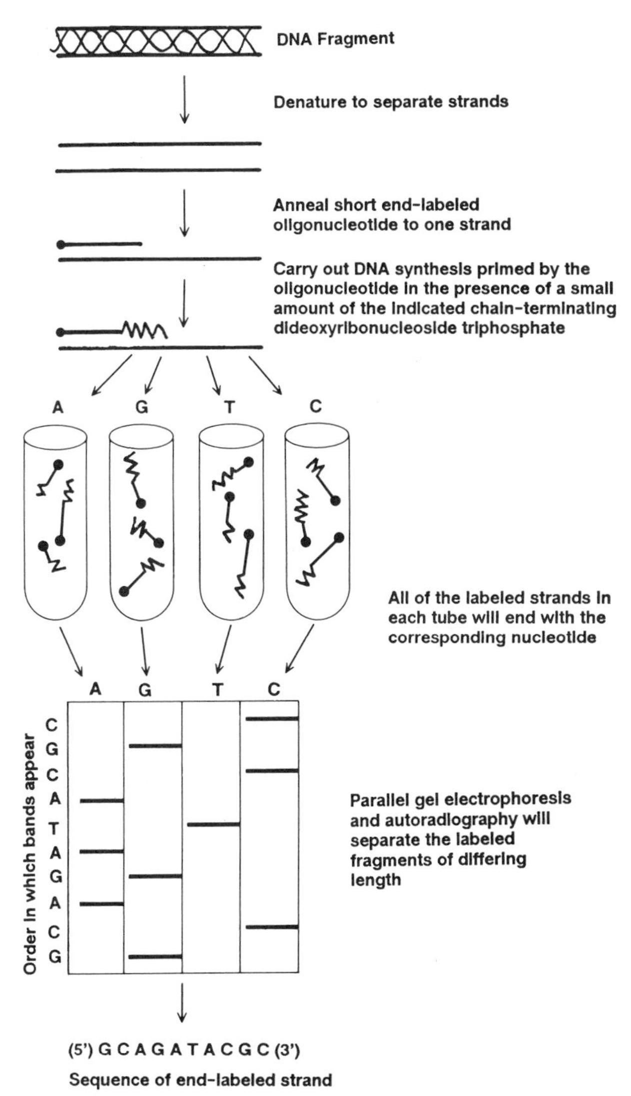


FIGURE 5-1 DNA sequencing by the enzymatic method. The key to this method is the use of a dideoxyribonucleoside triphosphate that blocks the addition of the next nucleotide after its incorporation into the growing chain. The primed in vitro synthesis of DNA molecules in the presence of a minor proportion of a single-type of such a chain-terminating nucleotide generates a family of DNA fragments each of which ends in the particular chain-terminating nucleotide (see also Figure 5-3). Here a radioactive DNA primer is used to initiate the synthesis of such DNA fragments and four different synthesis reactions—each with a different chain-terminating nucleotide—are analyzed by electrophoresis in four parallel lanes of a gel. The DNA sequence is then determined from the electrophoresis pattern.

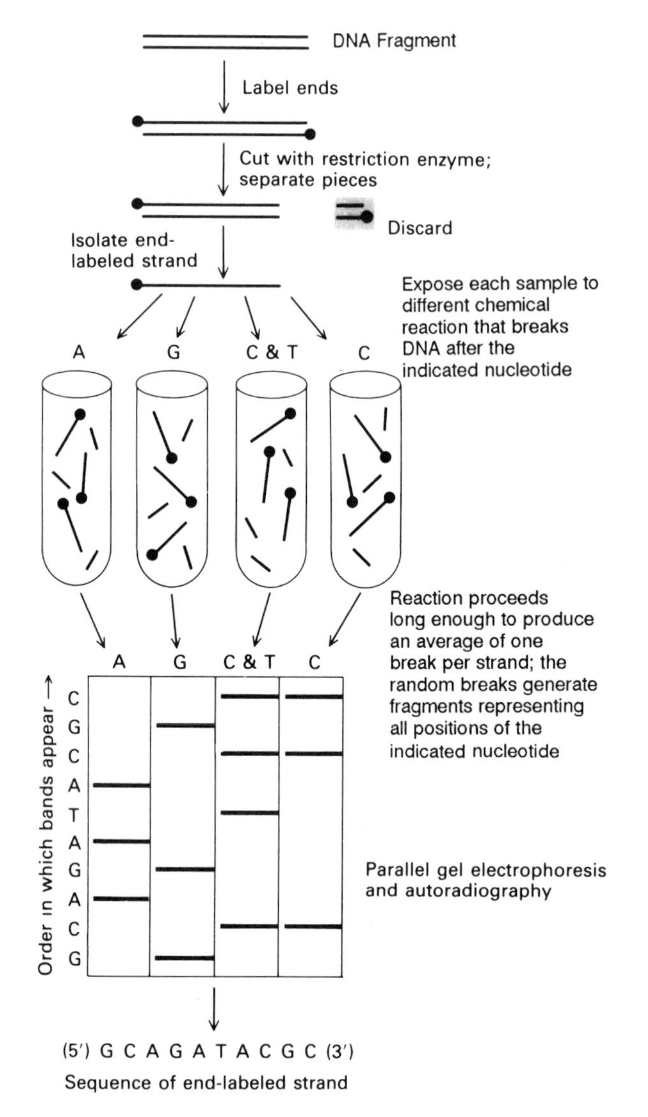


FIGURE 5-2 DNA sequencing by the chemical method. A DNA fragment that is radioactive only at its 5′ end is the material to be sequenced. A different chemical reaction in each of four samples breaks the DNA fragment only (or mainly) at A, G, both C and T, and C residues, respectively. The labeled DNA subfragments created by these reactions all have the label at one end and the cleavage point at the other. Electrophoresis of each sample through a polyacrylamide gel then allows each DNA subfragment to be separated according to its size. After autoradiography of the gel, the four sets of labeled subfragments (one set per gel lane) together yield a radioactive band for each nucleotide in the original DNA fragment. Adapted, with permission, from Darnell *et al.* (1986).

contiguous segment of human DNA determined to date is the 150,000 nucleotides encoding the human growth-hormone gene. This is 0.005 percent of the total genome.

Some other numbers are informative in this context. Currently, a skilled laboratory worker in a well-equipped facility can produce from about 50,000 nucleotides of finished DNA sequence per year (B. Barrell, Medical Research Council, Cambridge, personal communication, 1987) to about 100,000 nucleotides of finished sequence per year (E. Chen, Genentech, personal communication, 1987). The cost of this sequence ranges from $1 to $2 per nucleotide, an estimate based on the assumption that one worker costs a laboratory approximately $100,000 per year, including salary, supplies, and overhead. Even at the upper estimate of 100,000 nucleotides sequenced per person per year (which has not yet been achieved in a sustained effort), determining the human genome sequence would require 30,000 person-years of work at a cost of $3 billion. Since the sequencing of the genomes of other species is essential for an understanding of the human genome, the actual amount of sequencing would approach 6 billion nucleotides, at a current cost of $6 billion. This high cost of sequencing reflects the fact that the endeavor is still highly labor intensive and does not include unforeseeable technical problems or technical improvements.

Most of the time spent in a sequencing project is occupied with obtaining the original DNA clones containing the gene of interest and subcloning and handling the DNA prior to performing the actual sequencing reactions—steps that have not yet been streamlined or automated. In addition, the entire process from subcloning to interpreting gels requires careful supervision of personnel; a ratio of no more than three technicians for each doctoral-level scientist is generally accepted as optimal.

The rate of DNA sequence determination is also limited by the fact that all techniques currently use polyacrylamide gels that resolve no more than 250 to 500 nucleotides at a time. At this level of resolution, hundreds of millions of individual sequence determinations would be required to complete the human genome, given the estimate that each sequence will need to be determined three times over. By increasing the length of the average contiguous sequence that can be determined on a single gel, considerable time and effort would be saved.

## THE ACCURACY OF DNA SEQUENCING

Unless the human genome sequence is determined accurately, it will be of little use. Errors in DNA sequence determination occur at several levels. The most common is caused by insufficient resolution of adjacent DNA fragments in gel electrophoresis because of compression in their migration (neighboring bands merge into one). These effects are especially prevalent in regions containing large numbers of G and C nucleotides. Aberrations in the sequencing reactions can also occur in stretches of unusual sequence. These problems are compounded by human error, such as when researchers attempt to guess the sequence in ambiguous regions and when sequence gels are read past the point of accurate resolution. Another common source of human error occurs in transcribing the data into the computer. One potential source of error that will become more common as large-scale sequencing is attempted resides in the presence of short, highly repetitive sequences in human DNA, which can be confused when they occur in multiple clones. Furthermore, the cloning process itself may introduce a few errors.

The accuracy of DNA sequencing has not yet been firmly established. A careful and experienced laboratory probably achieves an accuracy of about one error in every 5,000 nucleotides (0.02 percent error rate) in the finished DNA sequence, but this degree of precision requires careful attention to virtually every nucleotide in the sequence (E. Chen, Genentech, personal communication, 1987). Such attention inevitably slows the sequencing rate. It will be difficult to hold the error rate to this level in a large-scale nucleotide sequencing project.

Although a few investigators have achieved a 0.02 percent error rate, most careful workers can only achieve an error rate of 0.1 percent. It is important to consider the impact of this error rate in the sequence of the human genome. Although it might seem large, the committee believes it is tolerable. The estimated level of DNA sequence heterozygosity among individuals is about 1.0 percent. The errors in the DNA sequence will be randomly placed, and hence most will occur outside coding sequences. Those errors in coding regions of genes that are either insertions or deletions of nucleotides (as most sequencing errors are) will have profound effects in that they will cause the reading frame to shift. This could lead to a failure to identify an exon as coding for part of a protein. If we consider that the average coding region (exon) is approximately 200 nucleotides long, one can anticipate that an error will occur on average in one of every five exons. The detection of some of these errors in exons may be facilitated by computer programs that predict coding regions on the basis of the use of particular sets of three nucleotides (codons) that code for each amino acid in humans. However, the errors will eventually be identified with certainty only by those interested in that region of the genome. This analysis puts into perspective the need to aim for approximately 0.1 percent as the maximum acceptable error rate in the initial sequence produced.

## EMERGING AND FUTURE TECHNOLOGY

The obvious mismatch between the efficiency of current DNA sequencing technology and the genetic complexity of genomes in even the simplest cells has given rise to several research projects aimed at developing more efficient sequencing methods. We seem to be on the threshold of a new generation of sequencing methods that should make large-scale sequencing projects more practical. Given the emergent state of these technologies, however, it is not surprising that expert opinion is widely divided on several key questions.

- Which of several next-generation strategies will prove most effective?
- Will the best next-generation strategy represent a quantum jump in sequencing capability or an incremental improvement that largely decreases the tedium of sequencing and shifts costs from skilled labor to instruments?
- In looking ahead to the need for a series of cumulative 5- to 10-fold increases in sequencing capability, is the future likely to lie in scaling up automated techniques that are already at the prototype stage, or does it lie in revolutionary new methods?

These questions will remain unanswered until future large-scale projects have been completed. Particularly crucial will be a determination of the steps in a sequencing project that become rate-limiting as the goals of sequencing are increased. No foreseeable technology will be able to automate DNA sequencing comprehensively. DNA sequencing involves a complex series of experimental steps with very different prospects for automation. For this reason, a given incremental increase in efficiency at any one step will rarely result in a comparable increase in overall efficiency.

Several current research projects aimed at automating various steps of sequencing are at different stages of development, and they illustrate the range of approaches being tested. They not only call for different technical strategies, but to various degrees they also reflect different perceptions of the steps in DNA sequencing most in need of greater efficiency. Several groups [California Institute of Technology, DuPont, and the European Molecular Biology Laboratory (EMBL)] are adapting the basic enzymatic sequencing methodology to a more automated operation at the level of reading the sequencing gels. Others are developing automated film readers, which are less expensive and not limited by the slow rate of electrophoresis (Elder *et al.*, 1985). Radioactive labeling of the fragments has been replaced by the use of fluorescent tags, which can be detected in the gel by characteristic light emissions evoked by laser illumination. The Cal Tech and DuPont methods allow more efficient use of the polyacrylamide gel, since the four reaction mixtures representing the four DNA nucleotide precursors can be labeled with different tags and then mixed together before being fractionated on a single gel lane (Figure 5-3) (Smith *et al.*, 1986). Alternatively, the EMBL method uses a single fluorescent tag for all four nucleotides and runs four gel lanes, which are monitored simultaneously with radioactive tags. In both cases, fragments are detected as they migrate past the point of laser illumination at the bottom of the gel, which eliminates the need to expose, develop, and interpret x-ray film. In each case, multiple sequences can be simultaneously analyzed on a single gel. The immediate goal of these projects is to develop a commercial instrument capable of sequencing 15,000 nucleotides per day, starting from the appropriate reaction mixtures.

A second approach is being developed in Japan, with assistance from the government and an industrial consortium that includes the Hitachi, Fuji, and Seiko corporations. This attempt to improve DNA sequencing emphasizes robotics and automated processing of samples. The automated steps begin with subcloned DNA fragments and carry them through the sequencing reactions. One such prototype performs more than 30 steps in the complex set of reactions that are required in the chemical sequencing strategy. Each step is controlled by a microcomputer. The maximum daily output of a single instrument is a sequence of 5,000 nucleotides. Current work emphasizes the organization of the overall sequencing experiment into a production line. The goal of this approach is to establish an automated facility able to sequence 1 million nucleotides per day at a cost of approximately $0.20 per nucleotide (Wada, 1987). This cost does not include the substantial cost of preparing the DNA fragments to be sequenced. The production-line approach would feature both automated and manual steps, with those operations most amenable to mechanization automated.

A third approach, called multiplex sequencing, depends less on automation and more on increasing the amount of sequence data that can be obtained from one set of chemical sequencing reactions fractionated on a single sequencing gel. Each sample analyzed contains a mixture of 40 or more DNA samples, each of which has been marked with a unique short nucleotide sequence (an oligonucleotide sequence). After the normal chemical sequencing reactions have been completed, the unlabeled samples are fractionated on a standard sequencing gel, and the separated DNA fragments are transferred to a membrane on which the spatial patterns of the fragments formed during electrophoresis are preserved. The sequencing ladder for each individual sample can then be successively visualized by a series of DNA-DNA hybridization assays, each using a different radioactive oligonucleotide as a DNA probe that is specific for the reference end of one particular subclone in the mixture. In principle, if a dozen sets of 40 mixed samples each are subjected to this analysis on a single gel and each can generate 250 nucleotides of DNA sequence, then a sequence of 120,000 nucleotides can be derived from one set of chemical sequencing reactions by using sequential hybridization with the membrane produced by this method (G. Church, Harvard University, personal communication, 1987).

All these methods utilize the chemistry or enzymology of current sequencing procedures. An intriguing question is whether fundamentally more powerful technologies are likely to arise in the foreseeable future. Little research is being directed toward this problem. The most obvious possibilities for future sequencing techniques would be the use of sensitive physical methods such as mass or fluorescence spectroscopy, magnetic resonance detection, and electron microscopy. These might be used in combination with each other or with more conventional biochemical fractionation methods. The disparity between the capabilities of the current technology and the magnitude of the work required to sequence the human genome suggests that fundamentally different technologies deserve serious exploration.

## OPTIONS AND RECOMMENDATIONS

The committee considered three options regarding the initiation of human genome DNA sequencing. The first is to begin a large-scale initiative immediately in one or a few large centers devoted to DNA sequencing with current technology. This option might be expected to include the establishment of an independent institute whose goal would be the mapping and sequencing of the genome as quickly as possible. The second option is to make a strong commitment to develop better DNA cloning, sequencing, and data analysis technologies by supporting smaller scale pilot projects (e.g., sequencing 1 million nucleotides in 1 year), while allowing investigators to gain practical experience with larger scale sequencing. These improvements in current technology should be designed to reduce the cost and increase the efficiency of sequencing techniques. The third option is to make no special effort to sequence the human genome, but to

FIGURE 5-3 Partially automated DNA sequencing by the enzymatic method. The method shown here is similar to that used in Figure 5-1. The main difference is that a fluorescently labeled primer is used to initiate the synthesis of the DNA fragments, rather than a radioactively labeled one. The four sets of differently labeled fragments are analyzed by the fluorescence as they move along a single gel lane. Adapted, with permission, from Alberts *et al.* (1989).

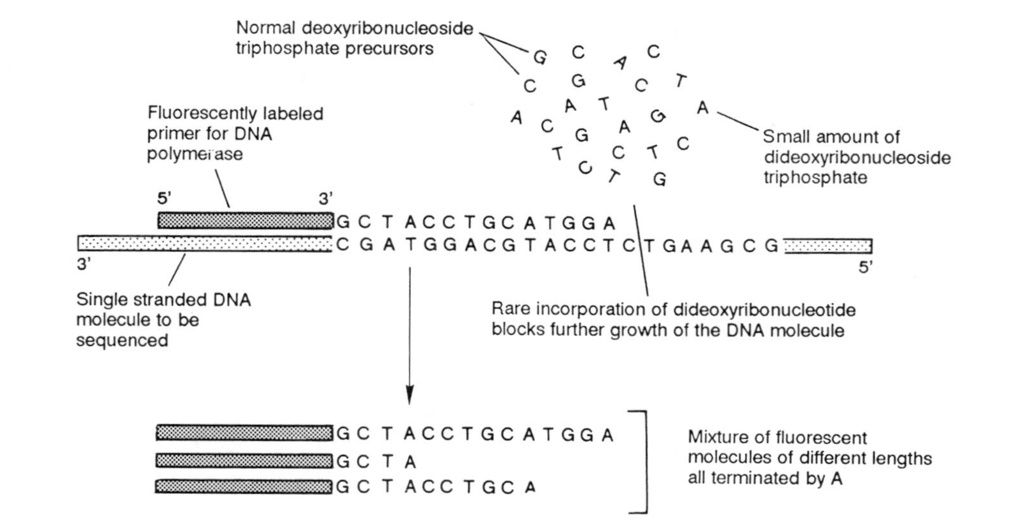

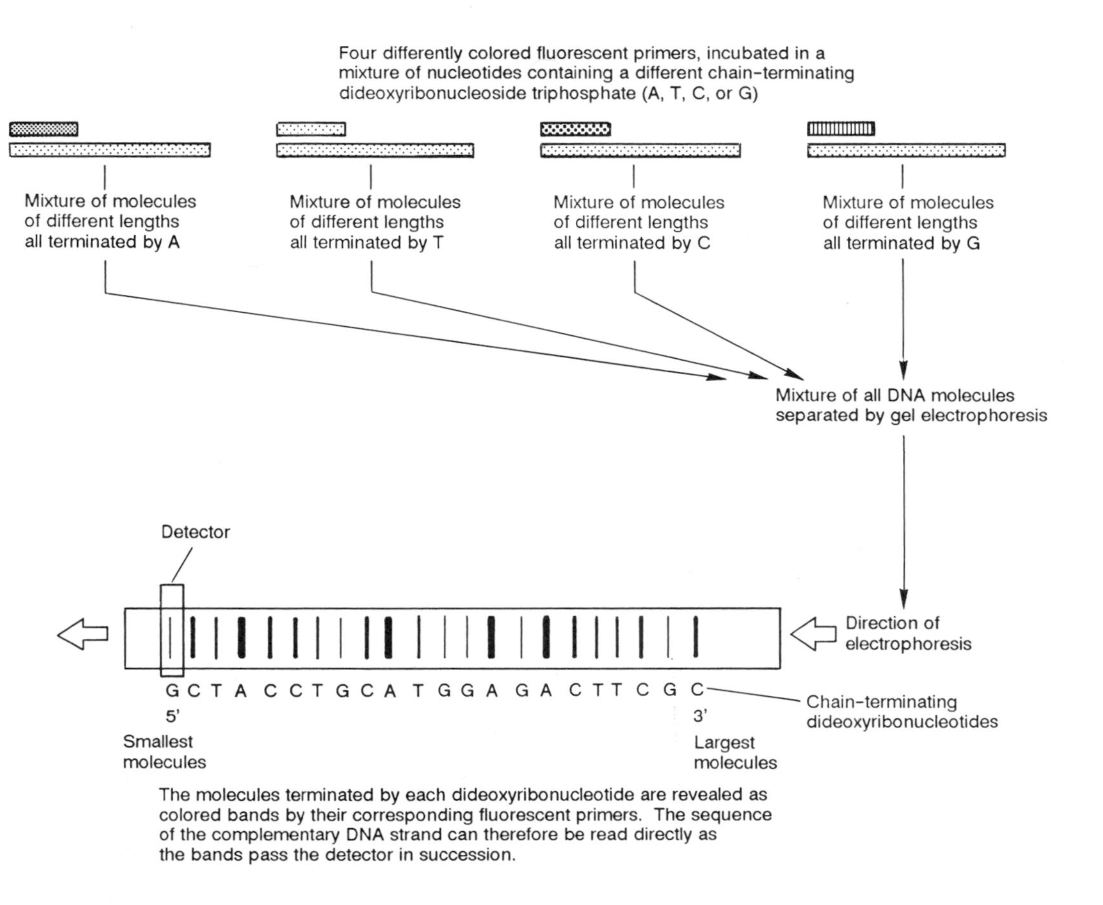
Four differently colored fluorescent primers, incubated in a mixture of nucleotides containing a different chain-terminating dideoxyribonucleoside triphosphate (A, T, C, or G)
Mixture of molecules of different lengths all terminated by A
Mixture of molecules of different lengths all terminated by T
Mixture of molecules of different lengths all terminated by C
Mixture of molecules of different lengths all terminated by G
Mixture of all DNA molecules separated by gel electrophoresis
Detector
Direction of electrophoresis
G C T A C C T G C A T G G A G A C T T C G C
5'
3'
Chain-terminating dideoxyribonucleotides
Smallest molecules
Largest molecules
The molecules terminated by each dideoxyribonucleotide are revealed as colored bands by their corresponding fluorescent primers. The sequence of the complementary DNA strand can therefore be read directly as the bands pass the detector in succession.

depend on the normal process of science to generate the sequence, knowing that the complete sequence would not be available within the next 20 years, if ever.

As explained in Chapter 3, knowledge of the sequence of the human genome and the genomes of the necessary reference organisms will provide a crucial medical and basic research tool that will be used by the biological and biomedical research community long into the future. The committee concluded that without a special effort to achieve this goal, the desired DNA sequences are not likely to be obtained in the time optimal for future medical and scientific advances, if ever. On the basis of this argument, the committee rejected option 3. In deciding between options 1 and 2, the committee concluded that the high cost and slow rate of sequencing with current technology precluded the initiation of a large-scale sequencing effort at the present time. Therefore, the committee made the following recommendations.

### *The Project Should Begin with Two Kinds of Studies*

Initially, improvements in existing technology and the development of new technology directed toward the long-range goal of a complete human genome sequence should be vigorously encouraged. This effort would include applications of automation and robotics at all steps in cloning and sequencing. It is particularly important to automate the steps of DNA cloning. In this context it is useful to think in terms of trying to achieve 5- to 10-fold incremental improvements in the cost, efficiency, or human labor required for these tasks. Several such incremental improvements are needed to make the sequencing of many important genomes practicable. A reasonable baseline sequence from which to measure initial progress is 150,000 nucleotides, the size of the largest human sequencing project to date.

These technological projects will assist in identifying the rate-limiting step in large-scale sequencing, which at present is believed to be the subcloning step—the one step that has not been automated. However, further improvements in all stages of the procedure from subcloning to the interpretation of sequence data will be required.

The awarding of competitive grants to individuals and to larger groups organized into cooperative, multidisciplinary centers is viewed by the committee as the most effective way to achieve these goals.

A second type of pilot study that should be initiated immediately would define as its goal sequencing approximately 1 million nucleotides of continuous sequence (approximately 5 to 10 times what has been achieved to date). Such projects will be important in providing an opportunity for the direct implementation and testing of improvements in existing technology as they arise and the provision of a practical impetus for the development of new technology. They will also reveal where problems in analysis are likely to arise. For example, will repetitive sequences complicate the assembly of a unique, contiguous sequence? Are some sequences unclonable? Will new genes be identified correctly?

As in the past, human gene sequencing by individual research groups interested in specific genes should be strongly supported by standard research grants. This directed sequencing will provide valuable information about genes that have been identified as important in biology and medicine and should also lead to advances in sequencing technology. However, as the physical map develops and as the cost and efficiency of DNA sequencing improve, ever-larger sequencing efforts taken on by groups interested primarily in the sequence of the genome as a goal in and of itself will evolve.

### To Derive the Full Benefit of the Human Genome Sequence Will Require Many New Tools, Including a Comprehensive Database of DNA Sequences from Other Organisms

Comparative sequence analysis has proven a powerful technique for distinguishing those elements of a gene sequence that are highly constrained functionally from those that are not. As explained previously, such analysis can provide insights into conserved regulatory and structural sequences. The availability of extensive sequence data from other organisms will also maximize the likelihood that the counterparts of important human genes will be identified in other organisms where their functions will generally be easier to study. The corollary will also hold: Genes that have been identified as important to other organisms will be found rapidly in the human DNA sequence. Therefore, a project of this type must not be restricted to determining the human genome sequence, but should include genome sequence determination of selected other species as well.

### DNA Sequence Determinations Require Quality Control

A mechanism of quality control must be developed to monitor the groups that are contributing extensive sequence DNA information. One might consider an external group that functions as the Bureau of Standards does to provide independent quality control. Quality control is critically important to the initiative, and it poses unique technical challenges. The optimum methods of checking DNA sequences are likely to differ from the optimum methods of collecting data; indeed, the sequence-checking method should ideally be experimentally independent of the sequencing method. For example, the presence of the many restriction enzyme cleavage sites predicted from the DNA sequence could be tested by cleavage of the DNA followed by gel electrophoresis.

To succeed, this project will require careful oversight and coordination among the groups involved in mapping, sequencing, collecting and analyzing data, and a system for distributing samples.

## REFERENCES

Alberts, B., D. Bray, J. Lewis, M. Raff, K. Roberts, and J. D. Watson. 1989. Molecular Biology of the Cell, 2nd ed. Garland, New York. In press.

Botstein, D., R. L. White, M. Skolnick, and R. W. Davis. 1980. Construction of a genetic linkage map in man using restriction fragment length polymorphisms. Am. J. Hum. Genet. 32:314–331.

Darnell, J. H. Lodish, and D. Baltimore. 1986. Molecular Cell Biology. Scientific American Books, New York. 1160 pp.

Doolittle, R. F., D. F. Feng, M. S. Johnson, and M. A. McClure. 1986. Relationships of human protein sequences to those of other organisms. Cold Spring Harbor Symp. Quant. Biol. 51:447–455.

Elder, J. K., D. K. Green, E. M. Southern. 1986. Automatic reading of DNA sequencing gel autoradiographs using a large format digital scanner. Nucleic Acids Res. 14:417–424.

Gilbert, W. 1978. Why genes in pieces? Nature 271:501.

Gilbert, W. 1985. Genes-in-pieces revisited. Science 228:823–824.

Gusella, J. F., R. E. Tanzi, M. A. Anderson, W. Hobbs, K. Gibbons, R. Raschtchian, T. C. Gilliam, M. R. Wallace, N. S. Wexler, P. M. Conneally. 1984. DNA markers for nervous system diseases. Science 225:1320–1326.

Maxam, A. M., and W. Gilbert. 1977. A new method for sequencing DNA. Proc. Natl. Acad. Sci. U.S.A. 74:560–564.

Sanger, F., S. Nicklen, and A. R. Coulson. 1977. DNA sequencing with chain-terminating inhibitors. Proc. Natl. Acad. Sci. U.S.A. 74:5463–5467.

Sanger, F., A. R. Coulson, G. F. Hong, D. F. Hill, G. B. Petersen. 1982. Nucleotide sequence of bacteriophage λ DNA. J. Mol. Biol. 162:729–773.

Smith, L. M., J. Z. Sanders, R. J. Kaiser, P. Hughes, C. Dodd, C. R. Connell, C. Heiner, S. B. H. Kent, and L. E. Hood. 1986. Fluorescence detection in automated DNA sequence analysis. Nature 321:674–679.

Staden R., and A. D. McLachlan. 1982. Codon preference and its use in identifying protein coding regions in long DNA sequences. Nucleic Acids Res. 10:141–156.

Wada, A. 1987. Automated high-speed DNA sequencing. Nature 325:771–772.

Wu, R., and E. Taylor. 1971. Nucleotide sequence analysis of DNA. II. Complete nucleotide sequence of the cohesive ends of bacteriophage λ DNA. J. Mol. Biol. 57:491–511.

# 6

# The Collection, Analysis, and Distribution of Information and Materials

---

The mapping and sequencing effort will generate more data than any other single project in the history of biology. For example, just to record the 3 billion nucleotides that make up the haploid human genome will require nearly 1 million pages of printed text. Variation between the two chromosomes of each individual (heterozygosity) and among the many human beings (polymorphism) further increase the body of information to be stored, collated, and analyzed. In the conception and planning of any human genome project, close attention must be paid to how the data and experimental material are collected and distributed.

The full set of information to be gained from mapping and sequencing the human genome is of potentially greater usefulness than its component parts. For example, although in principle one can use computer searches to pick out coding sequences that are parts of genes, finding the true beginning and end of a gene and all of its coding and noncoding components may require reference to other data, such as the similar nucleotide sequence from a related, but evolutionarily separated, species such as the mouse. To extract the maximum information from the human sequence, it will also be necessary to search for amino acid homologies with the entire set of all known proteins, regardless of their origin. In addition, extensive searches for regions of similarity of short nucleotide sequences between human genes and their mouse counterparts will be necessary to detect regulatory DNA sequences and other conserved sequences for which functions can then be sought. Finally, the correlation of sequence data with the large amounts of information derived from human genetic linkage and disease studies is needed to derive the molecular basis for human phenotypes. As more DNA sequence information is obtained, our sophistication in interpretation should increase to the point at which a computer search will reveal a fascinating wealth of correlative data concerning almost any new DNA sequence obtained.

The human genome project will differ from traditional biological research in its greater requirement for sharing materials among laboratories. For example, many laboratories will contribute DNA clones to an ordered DNA clone collection. These clones must be centrally indexed. Free access to the collected clones will be necessary for other laboratories engaged in mapping efforts and will help to prevent a needless duplication of effort. Such clones will also provide a source of DNA to be sequenced as well as many DNA probes for researchers seeking human disease genes. Two different types of centralized facilities will be needed: centers to collect and distribute materials such as DNA clones and human cell lines and centers to collect and distribute mapping and sequencing data.

The magnitude of the required data storage and distribution effort can be understood by comparing the existing content of facilities that collect and store mapping and sequence data with the anticipated capacity required if the human and other complex genomes are sequenced. For example, the DNA data bank maintained by GenBank and the European Molecular Biology Laboratory (EMBL) contains 15 million nucleotides of sequence data from the entire biological world—viruses, procaryotes, plants, and animals—and includes about 2 million nucleotides of human sequence. The human genome, containing 3 billion nucleotide pairs, is 200 times as large as the sum of these DNA sequences collected from all organisms. Moreover, only a few hundred restriction fragment length polymorphisms (RFLPs) have been mapped on the human genome, whereas the target of the genome project is several thousand mapped RFLPs—an increase of more than an order of magnitude. The efficient cataloging, management, and distribution of mapping and sequencing data at levels from one to two orders of magnitude greater than today's must be achieved in pace with data acquisition and are essential for the success of the project. Fortunately, several prototypic operations are already in place. These include GenBank/EMBL, Mendelian Inheritance in Man, Human Gene Mapping Library, and Centre d'Etude du Polymorphisme Humain, each of which is briefly reviewed below to provide a background for discussion of the much larger efforts that will be needed in the future. There are also repositories of cell lines and cultures, such as the American Type Culture Collection and the Cell

Bank in Camden, New Jersey, that have had extensive experience in handling and distributing biological materials. Although these operations are not reviewed here, they should be considered in the development of any materials-handling center.

## PRESENT INFORMATION-HANDLING ORGANIZATIONS

### *GenBank/EMBL*

The GenBank/EMBL data bank stores and distributes DNA sequence information. GenBank in the United States and the EMBL data bank in the Federal Republic of Germany share the task of recording, annotating, and distributing all the DNA sequence data published, regardless of the species of origin. Each bank is responsible for monitoring approximately half of the relevant published literature, and once they have entered and annotated the files, they exchange information so that each has a complete holding. The current holdings are about 15 million nucleotides, and the rate of acquisition is currently 7 million nucleotides per year and increasing rapidly (H. Bilofsky, personal communication, 1987). Both banks are finding it more and more difficult to keep up to date. The backlogs in the entry of published sequence data, a source of frustration within the user community, have two main causes.

First, nonelectronic data entry (entry from printed DNA sequences) still accounts for more than half of all data entered as a result of policy and organization, not technology. Authors have yet to be educated about the need to send data either electronically or in magnetic form to the data bases, in part because coordination between scientific journals and the data bases has, until very recently, been nonexistent. The reentry of data from a printed copy of a sequence into a data base is a slow, error-prone process, but in the absence of pressure from journals to authors to provide all sequence data in magnetic form, it has been absolutely necessary.

Second, GenBank/EMBL have not had sufficient support to keep abreast of the gene sequence data being generated by present biomedical research. However, the lessons that have been learned from their experience should be invaluable in setting up and managing a new facility dedicated to the collection of DNA sequence data, which will be an essential component of a human genome project.

### *Mendelian Inheritance in Man*

The Mendelian Inheritance in Man (MIM) project stores and classifies information about human disease phenotypes. MIM is an encyclopedia of gene loci based on human phenotypes, most of them disease phenotypes. It has been maintained at The Johns Hopkins University by Victor McKusick since the early 1960s and has been computerized since 1964. Seven hard-copy editions, all generated from the computer, have appeared between 1966 and 1986, and the number of entries has grown during this time from 1,500 to more than 4,100.

An attempt is made to assign only one entry per genetic locus; i.e., various phenotypes produced by alleles at one and the same locus (e.g., the beta-globin locus) are allowed only one entry. Inevitably, however, more than one entry has been assigned to many allelic disorders because of the incomplete status of our knowledge; in other cases a disorder assigned one entry subsequently proved to be produced by any one of two or more loci. Entries have also been created in the catalog for loci for which no Mendelian variation has yet been identified. Most of these are structural genes that have been characterized and mapped by a combination of somatic cell and molecular genetic methods.

Collaborative research in the management of this knowledge base at the Lister Hill National Center for Biomedical Communications of the National Library of Medicine has produced OMIM—an on-line version of MIM that is being tested in the clinic and laboratory. OMIM is designed to permit free movement between the text of MIM, gene map information, and a molecular defects list.

### *Human Gene Mapping Library*

The Human Gene Mapping Library (HGML) at Yale University positions genes and DNA landmarks on chromosomes (publication of Howard Hughes Medical Institute, 1986). The HGML consists of a number of separate but interrelated data bases. One of them, the "Map" data base, keeps track of the chromosomal positions of all mapped genes (currently more than 1,200). This is a dynamic data base: New genes are being entered at an accelerating rate, refinements of previous assignments are continually made, and the relations between gene map positions frequently change. The management of this data base requires constant attention to data input, editorial checking on the validity of the data, and data distribution. The data are maintained with an advanced data-base management system that is operated by a high-speed, large-volume computer. User-friendly menus have been prepared to facilitate access to the data by the inexperienced.

Other data bases within HGML include "Lit," which contains a list of all pertinent literature citations; "RFLP," which contains data on RFLPs; "Probe," which contains data on DNA probes used for mapping; and "Source," which contains information regarding the laboratories from which certain DNA probes or cell lines may be obtained.

The HGML data base, together with the scientific community it serves, also strives to maintain a uniform and orderly nomenclature for all mapped genes. It will be important to extend this nomenclature (or another that is agreeable to the scientific community) to other species, such as the laboratory mouse, so that direct comparisons between homologous genes in different species can be made readily. The HGML also assigns accession numbers to all DNA probes that might be useful as genetic markers. Upon request, researchers active in this field are given unique DNA probe identification numbers (D numbers), so that these probes can be described unambiguously. More than 2,000 probes have been numbered, and rapid growth to more than 100,000 is expected in the years ahead. An extension of this type of system could serve as a logical means of keeping track of the overlapping DNA clones produced by a human genome project.

### *Centre d'Etude du Polymorphisme Humain*

The Centre d'Etude du Polymorphisme Humain (CEPH) coordinates an international RFLP mapping effort using data from standard families (Marx, 1985; Dausset, 1986). CEPH, created by Jean Dausset in Paris, differs from MIM and HGML in being a collaborative research effort that both generates and stores human mapping data. As discussed in Chapter 4, CEPH maintains lymphoblastic cell lines and sends samples of DNA from these cells to collaborators in Europe and North America. In return, the recipients agree to test their RFLP probes on all the so-called "informative" families for each probe (the families in which two alleles of the particular RFLP are present). Collaborating members of CEPH are required to submit to Paris all of their RFLP probe mapping data in a prescribed, uniform format. CEPH then maintains a common data base to which members of the project have rapid access, which thereby allows members to place their own RFLP probes on a common linkage map. Through this collaborative project, the work of several laboratories on different continents is coordinated toward a common goal, which can be achieved much more rapidly than it could be in any one laboratory alone.

## MAPPING DATA BASES REQUIRED FOR A HUMAN GENOME PROJECT

### *The Collaborative Facilities Needed To Generate an RFLP Map Must Be Expanded*

One of the early goals foreseen for the human genome project is an RFLP map in which the average separation of markers is 1 cM. CEPH provides an example of how international collaboration, involving both the exchange of materials (DNA samples) and data (each group's probe-mapping results), can be organized for the production of an RFLP map held in common. However, to achieve a 1-cM RFLP map in a timely fashion (5 to 10 years), either CEPH must be expanded substantially or a new organization must be created and modeled along its lines, with the following objectives:

- A significant increase in the number and diversity of origin of families.
- Identification of several thousand new RFLP probes and their use to screen the set of DNAs obtained from these families (requiring either more laboratories or the enlargement of existing ones).
- A major increase in DNA production facilities because of the increased number of families and RFLP probes to be used with these DNAs.

At present, CEPH maintains stable lymphoblastoid cell lines derived from each of its 600 participants. It grows batches of the cells, extracts the DNA, and distributes it. More than one center may have to be established to grow the cells and to produce and distribute DNA.

At present, the laboratories collaborating with CEPH do not have to make available to the project their RFLP probe DNAs; they need only provide the data obtained with them. This helps to make the CEPH collaboration successful by avoiding constraints that might otherwise restrain participation. In the future, however, rules concerning the general availability of RFLP probes will have to be decided within the context of a human genome project. If RFLP mapping is done under contract by commercial enterprises, some of which already have considerable experience in the field, the contract should stipulate that there be open access to all the probes that are developed.

### *All Human Map Data Should Be Accessible from a Single Data Base*

In the major mapping data base associated with the human genome project, it will be necessary to keep track of the map positions, literature references, and material distribution sources for all identified landmarks in the human genome, including the DNA clones in the ordered clone collection. This can best be accomplished by having a single centralized data base that is easily accessible to the entire scientific community. A large data facility will be needed to manage this information. Initially, this facility would be responsible for integrating all RFLP mapping and DNA clone data, which would include all the information now in the MIM and HGML data bases. Once a human genome project begins generating large amounts of data, the annual management costs of mapping data bases are likely to rise from the total of $800,000 currently spent by MIM and HGML to perhaps $5 million. Whether the mapping data bases that are unified by a single management organization should also be housed under one roof is an open question. During the first stages of the project, and as long as MIM and HGML are electronically linked, it may be more practical to leave them in different institutions.

### *A Material Collection and Distribution Facility for Ordered Sets of Cloned DNA Fragments Will Be an Important First Stage in Any Sequencing Project*

The representation of the physical map in a DNA clone collection is immediately useful in that DNA segments of unknown origin can be located on them either by hybridization or by fingerprinting methods. Ultimately, the best physical map is the complete set of all such materials along with the information data bases described above. A separate dedicated facility will be required if these materials are to be made readily accessible to the entire user community.

Maintaining a facility that collects, organizes, and distributes all the available DNA clones generated by mapping efforts will be a major task. Further study will be needed to determine exactly how this facility should operate. At one extreme, one could imagine that such a facility would merely store DNA clones received from all participating laboratories (as DNA, as bacterial viruses, or as yeast cells carrying artificial chromosomes), index them according to some reasonable plan, and then redistribute them for a standard fee in response to specific requests from scientists. Because of the very large number of clones expected in the collection (at least several hundred thousand versus the 42,000 items now at the American Type Culture Collection), this aspect of the task will require major organizational efforts like those of a large mail-order company. In addition, stocks will also have to be replenished at intervals to keep the collection adequately supplied with materials. Because of possible clone instabilities, both these regrown stocks and each new stock received will require checking (by restriction enzyme fingerprinting or some other high-resolution method) as a standard quality-control procedure.

An additional possible routine role for the central facility includes converting large human DNA fragments cloned as artificial chromosomes into more readily accessible bacterial virus or cosmid DNA clone collections. The facility could also take all the DNA clones that have been mapped elsewhere by a variety of different procedures and fingerprint them by a single method to provide a standard indexing procedure. One can also envision a central facility that would actually help with the mapping effort; this type of facility could establish a single standard protocol for characterizing each DNA clone (for example, a standard restriction enzyme fingerprinting method) and collect and analyze the data provided by each of the participating laboratories to search for new overlaps. At present, mapping methods are in a state of flux, and many competing approaches are being tried in different laboratories. Any mapping role for a central facility should therefore be delayed until a reasonable consensus can be reached on the best way to proceed.

The cost of constructing and operating a storage and distribution facility will be high. Estimates of as much as $250 million spent for 30 years of operation have been made once the full range of clones has been generated (Stevenson, 1987).

## A DNA SEQUENCE DATA BANK DEDICATED TO A HUMAN GENOME PROJECT

### *A Concerted Initiative Aimed at Determining the Sequence of the Human Genome Will Generate Large Amounts of DNA Sequence Data*

Not only will there eventually be many billion nucleotides of human DNA sequences, but also there will be large tracts of sequence from the mouse genome that can be used for comparisons between the two species. In early stages of the sequencing portion of the project, it is likely that the genomes of experimental model organisms such as *E. coli*, yeast, the nematode, and *Drosophila* will be completely sequenced. If the project is to succeed, all data on large amounts of contiguous DNA sequence should be collected and distributed by a dedicated DNA sequence bank.

Fortunately, the amount of data associated with a human genome project is well within present disk storage and computer hardware capacity. Many government agencies—as well as the business world—are storing and handling significantly larger volumes of data. The difficulties will be encountered in the entry and classification of the data and even more so in their analysis and distribution to the international scientific community. An important goal of the entire endeavor should be to make available the information in a form that will benefit a very large portion of the biomedical and basic research community as quickly as possible.

### *All Data Must Be Entered Electronically or Magnetically*

From the outset, all sequence data must be entered into the DNA sequence bank by electronic or magnetic means. Moreover, the human genome project can circumvent many of the problems experienced by GenBank/EMBL by establishing a standard features format implemented at the point of data collection with the intention of expediting data entry. For example, all submitted data blocks could be packaged with references by the sender to data source, DNA clone number, chromosome region, and other factors. Since most data will probably be sent from fewer than two dozen research laboratories, the chances of entering spurious data from inexperienced investigators will be low. Nonetheless, there must be standards that set a minimum length of contiguous sequence suitable for submission and ensure quality control with regard to the frequency of errors in the accepted sequence.

### *An Initial Analysis Should Be Performed by a Central Facility*

Not unexpectedly, many different points of view exist about how, where, and when the large amounts of data in the genome sequence ought to be processed. New computers are constantly appearing, and new strategies for using them are always evolving. The most important analyses will no doubt be done by people interested in specific types of proteins, regulatory sequences, evolutionary processes, and so forth. However, some analysis should also be performed at the central facility to help in classifying the data for future research. Exactly how the data are to be analyzed might be tied to the number of centers or laboratories collecting the data, the kinds of staffing provided at a central facility, and the scope of the immediate data dispersal, i.e., whether it is national or international.

### *An Example of an Initial Sequence Analysis*

The strategy of data analysis will have to evolve as data accumulate, but the primary question will always be whether a particular sequence is an important island of information or just part of a surrounding ocean of chaos. Accordingly, the incoming data might be screened for repeated sequences. Even the most interesting parts of the human genome—the 50,000 to 100,000 genes—are going to be redundant, inasmuch as there will be many large families of closely related genes. The central nervous system, for example, which may account for 40 percent of all genes, is almost certainly going to include many such families. Some type of screening can help catalog the incoming data from the start and determine where and how the data should be stored.

To encourage the timely submission of data, all data submitted by the sequencing centers should be rapidly returned to them in a processed form for inspection and verification; each center should also be kept aware of progress at the others.

### *Establishing an Efficient Computer Network*

Many possible computer arrangements would suffice for the jobs described. It seems reasonable, however, to begin the operation on a modest level, with the intention of scaling up over several years. For example, the operation could easily be initiated with local computers that are connected with the National Supercomputer Network. In this model, the data collected at a sequencing center would be fed into a local computer, checked and entered into a features table, and then transmitted over the Supercomputer Network (which is especially good for high-speed transmission of large amounts of data) to the central DNA sequence bank. There, an analytical facility, which would probably use parallel computers at some future date, could handle the early stages of data analysis. The screened data would then be transmitted back to the various collection centers for verification. Once verified, the data would become available to the scientific community, moving through the Supercomputer Network to various local distribution points.

### *The Need for Research on Data Analysis*

We are only at the beginning of learning how to use computers to interpret DNA sequence information. New ways of searching DNA sequences will need to be designed as we learn more about such subjects as the binding sites for gene regulatory proteins, the rules that regulate RNA splicing, RNA secondary structure, and the effects of specific amino acid replacements on protein folding. In the future, we can expect to learn a great deal more about genes from their sequences than is possible today. A human genome project should therefore encourage the activities of those who combine skills in computer programming and biology; these individuals will be needed to generate the DNA sequence search routines of greatest utility to the biological community.

### *The Estimated Cost*

Although it is difficult to predict the cost exactly, approximately $5 million per year might be set aside for the sequence information facility. The largest part of the costs related to information handling will undoubtedly be devoted to professional staffing. Beyond that, funds will also have to be made available to develop software and to provide education and training to ensure further innovations in computer use in biology.

It is essential to keep the conventional data bases, including GenBank/EMBL, fully operational for the next several years, in particular to ensure comprehensive collection of sequence data from nonhuman sources. However, the time will come when sequence data from all sources will have to be melded into a single large, efficient facility.

## CONCLUSIONS

More than any other part of the human genome initiative, the handling of information and material will require organization and standardization. A single unified policy must prevail if the information is to be accurately acquired, stored, analyzed, and distributed. There must be a central facility for tracking and distributing the experimental materials, and there must be a dedicated computer center for storing, checking, screening, and searching the sequence and mapping data. The establishment of these facilities will be critical and will require careful advance planning. The committee recommends a competition in which all interested groups submit detailed applications or pilot program trials.

## REFERENCES

Dausset, J. 1986. Le centre d'etude du polymorphisme humain. Presse Med. 15:1801–1802.

Marx, J. L 1985. Putting the human genome on the map. Science 239:150–151.

Regional Localizations of Genes and Genetic Markers to Chromosomes and Subregions of Chromosomes. 1986, Number 1, HGM8. Howard Hughes Medical Institute Human Gene Mapping Library, New Haven, Conn.

Stevenson, R. E. Cited by L. Roberts, 1987. Human genome: Question of cost. Science 237:1411–1412.

# 7

# Implementation and Management Strategies

Earlier chapters have outlined the scientific strategies that seem most reasonable for genome mapping and sequencing studies and have argued in favor of an intensive effort to characterize the human genome in detail. In those chapters the committee has discussed the technological advances required for completion of the different human genome maps desired: a genetic map with RFLP markers spaced at about 1-cM intervals, a physical map of expressed genes (cDNA map), an ordered DNA clone collection that covers the entire genome, and physical maps of increasing resolution culminating in the DNA sequence. However, such a large endeavor requires a degree of organization and coordination that has no precedent in the biological sciences. The committee has argued specifically against establishing at this time one or a few very large production centers to carry out this project. Yet a concerted approach to project management will be required, and many of the benefits of a singular effort will be lost if results and materials are not quickly and efficiently shared.

The committee's recommendations immediately raise a number of obvious questions, such as, How should such an effort be funded so that its quality and success are ensured? How can the efforts of the different laboratories best be coordinated? How can the scientific community guarantee ready accessibility to all the information and materials to be generated? This chapter deals chiefly with the committee's attempts to answer these important questions.

## FUNDING A HUMAN GENOME PROJECT

### *Projects with the Potential to Make Substantial Technological Improvements in Genome Analysis Should Receive Top Priority*

Mapping and sequencing the human genome poses a severe technological challenge to our present abilities. A complete restriction map has recently been produced for the genome of the bacterium *Escherichia coli*, which is only 1/640 the size of the human genome (Smith *et al.*, 1987, Kohara *et al.*, 1987). In addition, the human genome is 20,000 times as long as the longest continuous stretch of DNA sequence of 150,000 nucleotides thus far produced in a single laboratory project. Interpreting the large amounts of information produced by large-scale DNA sequencing will be an even more formidable task. Thus, the human genome project is more ambitious by orders of magnitude than any single biology project completed thus far. Technological advances are needed in many different areas, and determining the most efficient technology to obtain the desired knowledge requires further research.

In view of the current situation, the committee recommends the establishment of a competitive grant program specifically focused on improving in 5- to 10-fold increments the scale or efficiency of mapping and sequencing the human genome. These grants would be designed to support work that is more technologically oriented than most ongoing biological research. For example, a project that aims at sequencing a fragment containing a contiguous segment of a million nucleotides or more might qualify for support, as might a project aimed at developing and testing an entirely new approach to DNA cloning or sequencing. In contrast, a project that aims to use standard approaches to isolate and characterize a single interesting human gene of 10,000 nucleotides would be more appropriately funded elsewhere.

### *Both Small Research Laboratories and Larger Multidisciplinary Centers Should Be Encouraged*

Nearly all the major methodological breakthroughs that have driven the modern revolution in biology have come from the efforts of small research laboratories, frequently those led by young investigators in the early stages of their careers. This trend will no doubt continue, and small groups that are already active in developing or improving relevant technologies should be supported by this project, irrespective of the particular genome that they are characterizing.

There is also much to be said for the establishment of multidisciplinary centers, in which 3 to 10 research groups in a single facility, each with a different but related focus, share equipment and personnel and collaborate to accomplish a larger goal than any single group could readily achieve. For the most part, these should be located at universities or research institutes that can help provide the projects with a constant influx of new people and ideas. Efficient mapping requires the coordination of a large number of different experimental and computer techniques. It also demands flexibility in developing and incorporating new technologies. The methods will probably evolve very rapidly, and the strongest and most efficient mapping efforts will provide their first critical tests. For these reasons, a substantial portion of the human genome mapping effort should probably be organized into medium-sized research centers, each with ongoing activities in both development of techniques and actual mapping and each with a reasonable fraction of the various technologies in place or under development. Such centers can set the stage for subsequent large-scale DNA sequencing projects. The sequencing effort will benefit from close contact with mapping efforts, some of which provide the samples to sequence, while others provide the framework to organize the sequence data generated and to assist in its interpretation.

### *The Establishment of a Single Large Production Center Is Not Advisable at Present*

The committee believes that it would not be wise to confine an activity such as mapping or sequencing to a single, large center at present. The task needs to be organized and coordinated, but it does not need to take place in a single location or laboratory. Unlike many physics projects of this magnitude, this large-scale biology project can be subdivided along several different lines. For example, an individual chromosome could be mapped by one laboratory, but even this is not necessary since available methods permit one to work efficiently with just a section of a chromosome. Similarly, while restriction, genetic, and DNA-clone-collection mapping must be conducted in parallel, they do not necessarily need to be done by the same investigators, inasmuch as materials generated by one method can readily be transferred for use in other laboratories.

There are strong technical and intellectual advantages to dispersing much of the mapping efforts among medium-sized, multidisciplinary centers (each with perhaps 30 to 100 persons). If adequately funded, these units would be large enough to accommodate biologists, chemists, physicists, and engineers with diverse skills and backgrounds, thereby achieving the critical mass necessary for an effective approach.

At the same time, available resources should make it possible to establish several centers with the same goals, each competing with other centers to advance the technology. Such competition is healthy and productive, and it permits subsequent judgments to be made on relative quality, allowing additional resources to be directed to the most successful units. Finally, the dispersal of these technology centers into existing universities and research institutes will allow members of the center to work closely with a large number of other scientists. This interface should be of great value to both groups, and it will ensure that the human genome effort will have the strong support it needs from the scientific community at large.

***Decisions for Funding Should Be Made by Peer Review***

The committee envisions that funding would initially be offered in the form of grants awarded for 3- to 7-year periods. It is imperative that these grants be awarded solely on the basis of scientific merit, as judged by panels of peer reviewers selected for their judgment and scientific expertise. The committee specifically recommends the form of review that has been routine for many years at the National Institutes of Health (NIH), in which the reviewers meet to discuss and debate the merits and faults of each grant application. Each application is then ranked relative to the others by the assignment of a priority score, which is forwarded to the individual NIH institute that distributes the research funds. Generally, grants are awarded in the order of their priority scores, until the allotment of funds for that cycle is exhausted.

As technologies mature, production units, such as contract organizations or dedicated centers will be required in the human genome project. For example, RFLP mapping is already a relatively mature technology. An RFLP map at higher resolution can be attained mainly by applying current methods on a larger scale. Such endeavors are appropriately supported by contracts rather than by grants. The central facilities that collect and distribute information and materials should also be supported by contract. Such efforts should also be subject to continuing peer review, both for technical competence and to ensure continuing coordination with the overall effort to map and sequence the human genome.

If a human genome project is funded by several separate U.S. government agencies as well as by private funds, an effective reviewing body will be needed to avoid excessive duplication of effort and to oversee cooperation between research groups. The committee recommends that the same body also ensure a uniform standard of peer review.

### *The Human Genome Project Requires New and Distinctive Funding of About $200 Million per Year*

To create the multidisciplinary centers suggested, new laboratories will need to be built, equipped, and staffed. Universities cannot be expected to provide the necessary resources without a major new source of funding from outside. The estimated cost of an effective project to map and sequence the human genome is $200 million per year. (This level of funding would be reached only during the third year of the project, with the first 2 years having lower levels of funding to allow a scale-up to an effective project.)

The money might be spent roughly as follows. In the first 5 years of the project there might be about 10 medium-sized, multidisciplinary groups and many smaller research groups working, with perhaps half of the projected total of 1,200 individuals in the multidisciplinary groups. Each professional researcher costs about $100,000 annually in pay and support (the standard number used in the biotechnology industry); thus this cost would be an estimated $120 million annually. Construction and equipment might cost about $55 million per year, the stock center, the data management center, the quality control effort, and the Scientific Advisory Board (see below) might cost approximately an additional $25 million per year.

As the project proceeds, annual construction costs will decrease, but the number of individuals participating in the effort may increase to about 1,500.

The committee's possible scenario for the project divides the effort into three 5-year periods (I, II, and III). During each period, mapping and sequencing efforts five times as complex as the next lower numerical designation would be undertaken at constant cost, reflecting five-fold increments in technological sophistication. Several points should be stressed.

- Attaining two successive fivefold increases in the technology for mapping and sequencing is an ambitious undertaking; if it is to succeed, a major effort must be expended in developing the required technology.
- The cost for sequencing must include preparation of the DNA samples. New methods of DNA subcloning and processing will have to be developed (or present ones automated) to attain the desired costs.
- DNA clones from an ordered DNA clone collection will be sequenced, thereby producing large, contiguous stretches of DNA sequence that are immediately useful; isolating the last 10 to 15 percent of these clones to fill in gaps in the map may be as expensive as

isolating the clones that cover the initial 85 to 90 percent of the genome.

• An ambitious effort of this type will require the recruitment of scientists with extensive experience in mapping and sequencing. The multidisciplinary centers supported by the project will presumably play a key role in training new independent scientists who can participate. The committee believes that the training of young scientists in the development of technology as well as its applications would be of the major benefit to the biological community.

A major objective of the human genome project would be to achieve an annual sequencing capacity of 1 billion nucleotides through the combined efforts of a modest number of centers by the year 2000. Once this ambitious goal is reached, it would be realistic to complete the entire human genome sequence, and powerful comparative studies on human polymorphisms and evolution would become possible.

Funding of the human genome project must not be at the expense of currently funded biological research. The essential purpose of the human genome project is to provide a resource to be used by biomedical scientists to accelerate the understanding of human biology and the application of this knowledge to human health. It would therefore defeat the purpose of the project if biomedical research and training were weakened by diverting funds from individual research programs or training grants. Major advances being made in other aspects of biology are expected to form the scaffolding required for interpreting and utilizing information resulting from this project.

### *Mapping Efforts Should Be Accelerated and Coordinated*

Most biologists feel that mapping the human genome is a valuable and attainable objective. At the present rate this goal will not be reached for many years. The committee strongly believes that this effort should be accelerated. Funds should be invested in projects that increase our technological skills, but large-scale mapping should begin even with present-day techniques. It should be possible to complete an RFLP map within 5 years with the investment of more money and the encouragement of coordinated efforts. Several types of physical maps can also be completed in a similar length of time.

### *The Sequencing Effort Should Evolve and Grow with Time*

Most of the human genome will probably become available in the ordered DNA clone collection as a result of the combined efforts of several multidisciplinary centers. If the envisioned support is forthcoming, the ordered DNA clone collection could be completed within 10 years. In principle, sequencing major blocks of the human genome could begin as soon as any contiguous area of a chromosome has been accumulated in the ordered DNA clone collection. Decisions for a major push on bulk sequence data collection, as distinct from the envisioned pilot projects that push technology development, would depend on how fast the new sequencing technologies develop.

Several relatively small genomes should be sequenced early in the project. Such genomes include that of yeast (0.5 percent of the size of the human genome), the nematode *Caenorhabditis elegans* (2 percent of the human), and the fruit fly *Drosophila melanogaster* (3 percent of the human). It is important to give these projects high priority, since they deal with widely studied experimental organisms, which, along with the mouse, provide the most important of the many model systems that will be required for interpreting all the sequence data that will be collected on humans.

### *International Collaboration on the Project Is Desirable*

There is a strong tradition of international cooperation in the biological sciences that has greatly speeded the rate of scientific progress in the past. This tradition must continue in any human genome project. Some portion of research on the human genome is bound to be done outside the United States, mainly in Europe and Japan. One project already started in Japan is supported by major industrial companies, which intend to automate DNA sequencing at the rate of 1 million nucleotides per day. In Europe, developments in semiautomatic DNA sequencing at the European Molecular Biology Laboratory in Heidelberg have resembled those at the California Institute of Technology. Work on new methods for physically mapping complex genomes, including the human genome, is progressing in Cambridge and London. In addition, extensive research on human genetics is under way in several European countries. The reference set of 40 families collected by the Centre d'Etude du Polymorphisme Humain and used throughout the world for mapping RFLPs was established in Paris, from which reference cells and DNA are distributed worldwide. GenBank in the United States currently shares on a 50:50 basis the collection and entry of DNA sequence data with the EMBL Data Bank in Heidelberg.

These examples suggest that the United States does not and cannot expect to monopolize information and innovation in this field. Moreover, the initiation of a human genome project in the United States will probably not deter work in other countries, but rather will stimulate it. Given this assumption, the importance of past traditions, and the magnitude of the task of mapping and sequencing the entire human genome, every effort should be made to enhance the existing contacts between U.S. laboratories and those overseas, so as to speed the work. Indeed, we believe it will become necessary to have some major organized mechanism for international cooperation. In particular, its objective would be to collate data and ensure rapid accessibility to it, as well as to distribute materials, such as cloned DNA fragments.

## MANAGING A HUMAN GENOME PROJECT

The human genome project presents complexities of organization and management at both the scientific and policy levels. At the biological level, the project differs from conventional biological research in that it must be coordinated in terms of mapping, definition of overlapping cosmids, distribution of DNA clones, sequencing, technology development, and data base design. At the policy and funding level, a number of governmental and private foundations will probably be involved. For these reasons, it is imperative to design a management system that will provide oversight, coordination, review of progress, and forward planning. The committee was convinced of the need for strong leadership for the project, and presents three possible management plans. We also recognize that the management of the human genome project may need to evolve as the project evolves, as have management mechanisms for similar projects with federal research support.

### *Three Possible Organizational Plans*

Briefly summarized, each of the possible organizational plans (designated A, B, and C) includes a Scientific Advisory Board, but differs in the administrative and funding leadership. In plan A a single federal agency serves as the lead for the project. This agency, which would be assisted by a Scientific Advisory Board composed of experts who provide scientific advice for the project, would be responsible for all aspects of the project.

In plan B an Interagency Committee, consisting of representatives of the National Institutes of Health (NIH), the Department of Energy (DOE), the National Science Foundation (NSF), and other federal agencies interested in the project, would be responsible for all aspects of the project. The Interagency Committee would be assisted by a Scientific Advisory Board that would provide advice on the project, as in plan A.

In plan C, an Interagency Committee would have the ultimate responsibility for the coordination and funding of the activities to be supported, while a single agency would be responsible for the daily administration of the project. In this plan, a Scientific Advisory Board would provide advice to both the Interagency Committee and the administrative agency.

Each of these management plans has advantages and disadvantages. Although the primary charge to the committee was not to develop an organizational plan, and decisions on management organization include considerations outside the committee's areas of expertise, a majority of the committee members favors plan A, which therefore is presented in the greatest detail.

### *Organizational Plan A: A Lead Agency and a Scientific Advisory Board*

*The Human Genome Project Should Be Assigned to a Major Federal Agency* In this plan the human genome project would be sited within a federal agency as an independently funded endeavor. This would place both the responsibility and the operation of the project within a single unit. Already there are aspects of the human genome project being supported by three federal agencies, NIH, DOE, and NSF. Each of these agencies has the potential capability to provide an effective home for the project.

Although the committee did not feel it was its role to designate a lead agency, it did discuss some of the merits of each agency. The NIH is the major agency today supporting research on the structure and functioning of DNA. It is also the agency with the mandate to foster biomedical research in the United States. It has a long history of successful support of peer-reviewed extramural research, successful intramural laboratories on the NIH campus, and has been involved in the oversight of large projects in biology, such as the viral cancer program. Recently, the NIH has shown interest in establishing a human genome project. DOE has successfully managed many important projects for the physics community, has supported some life science programs, has extensive computer facilities for data management, and has expressed strong interest in overseeing the human genome project. The NSF has been involved in the development of technology and instrumentation relevant to the human genome project, in the general support of basic biological research, and has a well-established peer review system.

Locating the project within one agency does not mean that all the funding for the project would flow from it or that scientists associated with another agency would not be able to obtain funding. For example, even if DOE were not the lead agency, scientists at the national laboratories and DOE grantees should be eligible to compete for research support relevant to the human genome project. The same would apply for scientists at the NIH and NIH grantees, if NIH were not the lead agency.

*Scientific Advisory Board* Although the lead agency would have the ultimate authority and responsibility for the funding and administration of the project, our committee believes that the overall scientific oversight of the project should draw upon the experience and be guided by a Scientific Advisory Board (SAB) made up predominately of scientists with expertise in the methods and goals of the project, and chaired by a full-time chairman who is a distinguished scientist. We recommend that the members of the SAB be appointed by the lead agency for staggered terms. The chairman should be appointed for a fixed term with a possibility for reappointment and should be provided with a full-time staff sufficient to carry out the directives of the lead agency and the SAB. The role anticipated for the Board is somewhat stronger than that of a typical scientific advisory board.

The major responsibilities of the SAB include:

- To facilitate coordination of the efforts of the many laboratories that are expected to participate in this effort.
- To help assure the accessibility of all information and materials generated in the project by advising on the oversight of the data center and the stock center and recommending contracts where appropriate. It would oversee formation of standard terminologies and reporting formats so that the large body of information to be obtained can be readily communicated and analyzed by the entire scientific community.
- To monitor the quality of research by helping to assure a uniform standard of peer review.
- To suggest mechanisms for strict quality controls on the sequence and mapping data collected.
- To promote international cooperation, serving as a liaison to projects outside the United States regardless of their funding sources.
- To make recommendations concerning the establishment of large sequencing endeavors, thereby balancing focus with breadth.
- To publish periodic reports stating progress, problems, and recommendations for research.

*The Scientific Advisory Board Should Provide Advice on the Peer Review Process and on Coordination of the Project* A human genome project cannot succeed unless the various mapping and sequencing efforts are coordinated. Standards must be set and the rapid distribution of materials and information must be facilitated. The failure of a unit to coordinate its efforts with others should jeopardize the support of the unit not in compliance.

The lead agency and SAB should work closely in developing and implementing a high standard of peer review. After the committee considered several options, it concluded that this system would function best if the SAB were involved in monitoring the evaluation of proposals for funds. After consideration of the advice of the SAB, the lead agency would then fund selected applications.

*The Scientific Advisory Board Would Require Funding* To be effective, the SAB must be adequately funded. Although the lead agency would provide much of the money for the SAB, private foundations and institutes should be encouraged to help support it. An appropriate mechanism for merging private and federal funds for this purpose would have to be developed. If, in addition to its role as scientific advisor and coordinator, the SAB is assigned such tasks as oversight of peer review panels, funds will have to be provided to it for these purposes.

### *Organizational Plan B: An Interagency Committee and Scientific Advisory Board*

*The Interagency Committee* Three government agencies can potentially play leading roles in the human genome project: The National Institutes of Health (NIH) because of their central responsibility for human biomedical research and their exemplary peer-reviewed extramural grant programs; the Department of Energy (DOE) because of its interest in the project and experience in data management and the management of large-scale projects; and the National Science Foundation (NSF) because of its commitment to technology development and basic research support of biology across the discipline. Private foundations and institutes may also be interested in involvement in the project.

In this plan the above three governmental agencies would form an Interagency Committee (IAC) that also includes members from other agencies interested in the project. The chairmanship of the IAC might rotate among the three principal agencies (NIH, DOE, and NSF) involved in the project.

The IAC would be responsible for overall administration of the project, including funding of research programs and supporting services; administration of a common peer review process, the stock center, and the data center; and the appointment of and response to a Scientific Advisory Board (SAB). Because funding for this project may come from several agencies it is important that the Interagency Committee be responsible for the coordination of the funding. Each year the committee should develop a total project budget and determine what the contributions of each of the agencies would be to the project. Representatives of each agency can then request the funds required for the project from the administration and the respective appropriations committees.

The Scientific Advisory Board would assume a role similar to that outlined in organizational plan A, except that it would of course advise the Interagency Committee rather than a lead agency.

### *Organizational Plan C: Interagency Committee, Administrative Agency, and Scientific Advisory Board*

In this final proposed plan aspects of organizational plans A and B are combined to form a three-part administrative structure.

*The Interagency Committee* As discussed under organizational plan B, an Interagency Committee would be established to oversee the project. It would be ultimately responsible for the coordination and funding of the activities to be supported, the administration of the peer review process, the research program, the stock center, and the data center. It is expected that this committee, which would provide the administrative and the funding lead for the project, would pay close attention to the recommendations of the Scientific Advisory Board and the administrative agency.

*The Administrative Agency* In this plan, one of the member agencies of the Interagency Committee would act as the administrative agency responsible for the daily administration of the project. This agency would be involved in the administration of funds and other administrative aspects of the project, such as the operation of the stock center and the data center. Questions and inquiries about the project would be directed to this agency which would serve the important role of clearinghouse for the effort. This agency would help to guarantee that the project is well-run, and that the necessary details in the operation of this large-scale project are completed.

The administrative agency would work closely with the SAB in developing and implementing a high standard of peer review. This agency would arrange the administrative details of the peer review process, while the SAB would monitor the initial evaluation of grant and contract proposals. After the assignment of a priority number to each grant and contract, the Interagency Committee would select the applications to be funded.

The Scientific Advisory Board in this organizational plan would have similar responsibilities to those outlined in plan A, but would advise both the Interagency Committee and the administrative agency.

## REFERENCE

Kohara, Y., K. Akiyama, and K. Isono. 1987. The physical map of the whole *Escherichia coli* chromosome. Cell 50:495–508.

Smith, C. L., J. G. Econome, A. Schutt, S. Klco, and C. R. Cantor. 1987. A physical map of the *Escherichia coli* K12 genome. Science 236:1448–1453.

# 8

# Implications for Society

---

The applications and implications for biology and medicine of a project to map and sequence the human genome have been mentioned often in this report. In this final chapter we discuss some of the other issues for society, including the commercial, legal, and ethical implications of such a project.

## COMMERCIAL AND LEGAL IMPLICATIONS

Mapping and sequencing the human genome will result in new information and materials of potential commercial value, for example, clones that encode previously undiscovered hormones, growth factors, or mediators of immunity. The commercial value of these resources raises questions concerning possible copyright protection of the data and ownership of the intellectual property and materials generated by participants in the human genome project. Should it be possible to copyright sequences from the human genome and, if so, by whom? Should a central agency of the government own the patents for new materials, such as DNA clones generated by this project? What are the implications for international collaboration? Because these are complex issues requiring study by scientists, lawyers, and policy-makers, the committee believes that they should be given prompt study by an independent body. It is important to resolve the legal issues concerning the conduct of the human genome project. Absolutely essential to the success of the project will be cooperation between laboratories and centers—within the United States and internationally—and the ready availability of data and materials to all participants. This committee believes that human genome sequences should be a public trust and therefore should not be subject to copyright.

## ETHICAL AND SOCIAL IMPLICATIONS

Whatever its scientific merits, a concerted effort to map and sequence the human genome would have profound social significance. Human beings are fascinated with the reasons we are what we are, both for what those reasons tell us about ourselves and for the insights they give us into those around us. In this context, the prospect of a complete biological book on humankind provokes both excitement and concern and raises philosophical and ethical questions. Three sorts of questions seem particularly important to reflect upon in advance of any genome mapping and sequencing effort: How should the project proceed? How should the information be interpreted? To what use should the resulting information be applied? None of these are new questions for human geneticists. In fact, the ethical and social challenges presented by a human genome mapping and sequencing project are largely the same as those already addressed by scientists, clinicians, patients, and policymakers in other settings (Macklin, 1985). Still, the scale and significance of this project require that these questions be carefully assessed in this context.

### *Conducting a Genome Mapping and Sequencing Project*

The ethical considerations involved in conducting this project are shared by those conducting any biomedical analysis of human tissue. One consideration concerns privacy and confidentiality. The privacy and autonomy of the individuals who contribute the material studied must be protected. For most research in this project, this goal is easily accomplished: The isolated cell lines and genetic materials analyzed will come from a wide variety of sources, through standard channels designed to preserve the confidentiality of the contributors and ensure that their participation is voluntary (U.S. Congress, House Committee on Science and Technology, 1986). However, where family histories are studied to produce genetic linkage maps, geneticists will sometimes face ethical dilemmas over maintaining confidentiality or disclosing research findings to a relative discovered to be at risk for genetic disease. Again, this is not a new problem for human geneticists (Capron, 1979). As the mapping research proceeds, it will become increasingly important to reconfirm the geneticist's traditional willingness to take on the burden of responsibility in decisions to break confidentiality and to consider such a break only when the probability is high that serious, avoidable harm would otherwise come to identifiable individuals (President's Commission, 1983).

### *Interpreting the Medical Implications of Genetic Information*

Mapping and sequencing the human genome could provide a great deal of new knowledge about the genetic basis of human disease. However, the effects of that knowledge will be highly colored by the way its practical implications are interpreted. Without careful interpretation, information that links particular genes with disease can have harmful consequences for the people who carry those genes, quite apart from the disease itself.

For example, without clear guidance it would be easy for people to misinterpret statistical correlations between clinical diseases and particular genetic markers, so that they take the discovery of the marker to be diagnostic of the disease. Genetic susceptibilities, predispositions, or risks for disease are variable and sometimes ambiguous concepts (Lappe, 1979a). If interpreted too strongly, preventive efforts could force certain groups or individuals to assume the social and psychological burdens of the afflicted unnecessarily. For example, only 0.10 percent of those who have the HLA B 27 marker associated with ankylosing spondylitis will ever develop the disease (Lappe, 1979b). That association, however, could heighten the anxieties and affect the plans of many more people if it is misunderstood or overstressed.

These misinterpretations can also affect our social policies. Because of the connection we make between our genetic constitutions and our identities as individuals, diagnoses that trace diseases to our genes can also convey stigma and set the stage for social prejudice (Ablon, 1981). It will be the burden of the researchers to interpret the correlations they draw as clearly as possible, to avoid simplistic associations between genetic markers and clinical conditions, and to educate clinicians and the public about the actual implications of their findings for individuals.

Moreover, even where prognostic information about disease is interpreted correctly, it may still be clinically problematic. Where there is no effective therapy, new abilities to detect diseases in advance of their onset create harder choices for clinicians and patients. As we explore the human genome, more people will be faced with the dilemma that now faces those at risk from Huntington's disease: Is it better or not to know one's fate when it is out of one's control? At the same time, the very discoveries that exacerbate those dilemmas will also be crucial steps in developing of the new therapies that can help resolve them. It will be important as the project proceeds to pursue those steps and attempt to narrow, rather than widen, the gap between our abilities to diagnose and treat disease (Fletcher and Jonsen, 1984).

### *The Use and Abuse of a Complete Genome Map*

Probably the most contentious set of social problems resulting from a human genome project would be in the use of its findings. As a by-product of the project, a great number of new diagnostic tests for specific traits and conditions will become available. The scientific and medical communities will receive an increasing variety of screening requests, ranging between those from couples making reproductive decisions to those from employers planning personnel policies. The issues they will face in considering those requests again return to the very personal nature of the information the screening tests yield: Is it ever appropriate to screen an individual for the benefit or profit of some other person or institution?

The most controversial applications of the new genetic screen would be their use by industries and insurance companies to identify individuals who might be occupational or insurance risks (Murray, 1983). As the human genome project proceeds, the ongoing discussion of these practices, and the need for sound social policy about them, will only intensify. Questions about protecting individual autonomy, the ownership of genetic information, and the interpretation of map-based medical prognoses will figure heavily in this discussion. To a large extent, any changes in social policy will reflect the ways those same questions are addressed by the scientific community in conducting the project.

Ethical questions about the appropriate use of genetic information may also be raised within the more intimate circle of the nuclear family. For example, are there limits on the traits that parents may decide their children must have? Traditionally, these limits have been set at the boundaries of the pathological conditions; screening requests for traits that have no pathological import, such as the sex of the child, are usually denied (Juengst, 1987). Yet the boundaries of conditions that might be regarded as pathological are vague. As genetic markers become available for an increasing range of traits, the ability to identify those markers prenatally will present difficult decisions for clinical geneticists: What levels of disease susceptibility or risk warrant prenatal diagnosis? Are prenatal tests for somatically correctible genetic defects, diseases with late onset, or minor defects appropriate?

Once again, these questions are not unique to the effort to map and sequence the human genome. They are all questions already presented to clinicians, geneticists, and prospective parents by current diagnostic techniques. By making an increasingly wide range of screening tests available, however, the human genome project is likely to increase the frequency with which these questions arise and the need for settled professional and social approaches to them. Fortunately, in the development of social policy and professional ethics with regard to these questions, it is already possible to draw on the resources of a large literature base and lively public discussion (for example, see Milunsky and Annas, 1985). Important steps toward social consensus on the issues have even taken place at the national level. For example, the reports of the President's Commission for the Study of Ethical Problems in Medicine and Biomedical and Behavioral Research (1982, 1983) already offer a useful orientation that can help meet the ethical challenges that mapping and sequencing the human genome would present.

Finally, it should be noted that RFLPs will continue to be developed, maps will be made, and genetic counseling will occur even without a concerted effort to map and sequence the human genome. The greater coordination and quality control that will result from a concerted effort will in fact benefit the public by reducing the chance of misuse of poorly organized information.

## REFERENCES

Ablon, J. 1981. Stigmatized health conditions. Soc. Sci. Med. 15B:5–9.

Capron, A. M. 1979. Autonomy, confidentiality and quality care in genetic counseling. In A. M. Capron *et al.*, eds. Genetic Counseling: Facts, Values, and Norms (Birth Defects: Original Article Series, vol. 15). Alan R. Liss, New York. Pp. 307–340.

Fletcher, J., and A. Jonsen. 1984. Ethical considerations in prenatal diagnosis and treatment. In M. R. Harrison, M. S. Golbus, and R. A. Filly, eds. The Unborn Patient: Prenatal Diagnosis and Treatment. Grune and Stratton, New York. Pp. 159–167.

Juengst, E. 1987. Prenatal diagnosis and the ethics of uncertainty. In J. F. Monagle, and D. C. Thomasa, eds. Medical Ethics: A Guide for Health Care Professionals. Aspen, Rockville, Md. Pp.23–32.

Lappe, M. 1979a. Theories of genetic causation in human disease. In A. M. Capron *et al.*, eds. Genetic Counseling: Facts, Values, and Norms (Birth Defects: Original Article Series, volume 15). Alan R. Liss, New York. Pp. 3–47.

Lappe, M. 1979b. Genetic Politics: The Limits of Biological Control. Simon and Schuster, New York.

Macklin, R. 1985. Mapping the human genome: Problems of privacy and free choice. In

A. Milunsky and G. J. Annas, eds. Genetics and the Law III. Plenum, New York. Pp. 107–115

Milunsky, A., and G. J. Annas, eds. 1985. Genetics and the Law III. Plenum, New York.

Murray, T. H. 1983. Genetic screening in the workplace: Ethical issues. J. Occup. Med. 25:451–454.

President's Commission for the Study of Ethical Problems in Medicine and Biomedical and Behavioral Research. 1982. Splicing Life: The Social and Ethical Issues of Genetic Engineering with Human Beings. Government Printing Office, Washington, D.C.

President's Commission for the Study of Ethical Problems in Medicine and Biomedical and Behavioral Research. 1983. Screening and Counseling for Genetic Conditions: The Ethical, Social and Legal Implications of Genetic Screening, Counseling, and Education Programs. Government Printing Office, Washington, D.C.

U.S. Congress House Committee on Science and Technology, Subcommittee on Investigations and Oversight. 1986. The Use of Human Biological Materials in the Development of Biomedical Products. 99th Cong., 1st sess. Government Printing Office, Washington, D.C.

APPENDIX

# A
# Glossary

The following have been taken directly or modified from definitions in *A Dictionary of Genetics*, 3rd edition, by Robert C. King and William D. Stansfield, Oxford University Press, New York, 1985, with permission from the publisher.

**Allele** One of a series of possible alternative forms of a given gene, differing in DNA sequence and affecting the functioning of a single product (RNA and/or protein).

**cDNA** Complementary DNA produced from a RNA template by the action of RNA-dependent DNA polymerase (reverse transcriptase).

**Chromosome** (1) In prokaryotes, the circular DNA molecule containing the entire set of genetic instructions essential for life of the cell. (2) In the eukaryotic nucleus, one of the threadlike structures consisting of chromatin (DNA plus associated protein) and carrying genetic information arranged in linear sequence.

**Clone** (1) A group of genetically identical cells or organisms all descended from a single common ancestral cell or organism by mitosis in eukaryotes or by binary fission in prokaryotes. (2) Genetically engineered replicas of DNA sequences.

**Codon** The nucleotide triplet in messenger RNA that specifies the amino acid to be inserted in a specific position in the forming polypeptide during translation.

**Cosmid** Vectors designed for cloning large fragments of eukaryotic DNA.

**Crossing over** The exchange of genetic material between homologous chromosomes.

**Cytogenetics** The science that combines the methods and findings of cytology and genetics.

**Electrophoresis** The movement of the charged molecules in solution in an electrical field. The solution is generally held in a porous support medium, such as a gel made of agarose or polyacrylamide. Electrophoresis is generally used to separate molecules from a mixture on the basis of differences in net electrical charge and also by size or geometry of the molecules, in a manner that depends on the characteristics of the gel matrix.

**Exon** A portion of split gene that is included in the transcript of a gene and that survives splicing of the RNA in the cell nucleus to become part of a messenger RNA or a structural RNA in the cell cytoplasm.

**Gene** A hereditary unit that, in the classical sense, occupies a specific position (locus) within the genome or chromosome; a unit that has one or more specific effects upon the phenotype of the organism; a unit that can mutate to various allelic forms; a unit that codes for a single protein or functional RNA molecule.

**Intron** In split genes a segment that is transcribed into nuclear RNA, but is subsequently removed from within the transcript by RNA splicing and rapidly degraded. Most genes in the nuclei of eukaryotes contain introns.

**Linkage map** A chromosome map showing the relative positions of the known genes on the chromosomes of a given species, as determined by the inheritance of linked traits.

**Oligonucleotide** A polymer made up of a few (between 2 and 20) nucleotides.

**Open reading frame** Regions in a DNA molecule where successive nucleotide triplets can potentially be read as codons specifying amino acids and where the sequence of these triplets is not interrupted by stop codons.

**Polymorphism** The existence of two or more genetically different classes in the same interbreeding population (Rh-positive and Rh-negative humans, for example).

**Recombination** The occurrence of progeny with combinations of genes other than those that occurred in the parents as a result of independent assortment or crossing over.

**Restriction fragment length polymorphisms** Variations occurring within a species in the length of DNA fragments generated by a specific endonuclease. Such variations are generated by mutations that create or abolish recognition sites for these enzymes. For example, restriction endonuclease mapping of human structural genes for beta hemoglobin chains has shown that parents with the sickle cell mutation produce abnormal restriction fragments.

**Reverse transcription** DNA synthesis from an RNA template, mediated by reverse transcriptase.

**Somatic cell** Any cell of the eukaryotic body other than those destined to become sex cells. In diploid organisms, most somatic cells contain the $2N$ number of chromosomes.

**Stop codon** A ribonucleotide triplet signaling the termination of the translation of a protein chain.

APPENDIX

# B
# Curricula Vitae of Committee Members

---

**BRUCE ALBERTS** received a Ph.D. degree in biophysics from Harvard University and is a professor of biochemistry at the University of California, San Francisco. He is a member of the National Academy of Sciences and conducts research on the structure and function of multiprotein complexes and the chemistry of DNA replication.

**DAVID BOTSTEIN** holds a Ph.D. degree from the University of Michigan and serves as a professor of genetics at the Massachusetts Institute of Technology. He is a member of the National Academy of Sciences and conducts research on the genetics of the cytoskeleton and cell cycle in yeast, secretion of proteins in yeast and bacteria, and the use of DNA polymorphisms to construct linkage maps in humans.

**SYDNEY BRENNER** was educated at the University of the Witwatersrand and Oxford University (D.Phil.) and is now a member of the scientific staff of the Medical Research Council Laboratory of Molecular Biology and a fellow of Kings College, Cambridge. He is a fellow of the Royal Society, a foreign associate of the National Academy of Sciences, and a recipient of the Lasker award. His research interests include molecular biology of development and gene mapping and sequencing.

**CHARLES CANTOR** received a Ph.D. in physical chemistry from the University of California, Berkeley, and is now a professor and chairman of genetics and development at the Columbia University

College of Physicians and Surgeons. His research focuses on methods for handling very large nucleic acids and proteins and on the structure of complex nucleoproteins, such as chromosomes and viruses.

**RUSSELL DOOLITTLE** earned his Ph.D. degree from Harvard University and is now a professor of chemistry at the University of California, San Diego. A member of the National Academy of Sciences, he does research on the structure and function of fibrinogen and the evolution of proteins.

**LEROY HOOD** holds an M.D. degree from the Johns Hopkins University and a Ph.D. degree from the California Institute of Technology, where he is now a professor of biology. He is a member of the National Academy of Sciences and a recipient of the Lasker award. His principal research interests are in the molecular biology of the major histocompatibility complex and the T-cell receptor genes, as well as the development of instrumentation for molecular biology.

**VICTOR McKUSICK** earned an M.D. degree from the Johns Hopkins University, where he is now a professor of medical genetics. He was for 12 years chairman of the Department of Medicine and physician-in-chief of the Johns Hopkins Hospital. He is a member of the National Academy of Sciences and a fellow of the Royal College of Physicians (London). His research is concerned with human genetics.

**DANIEL NATHANS** received the M.D. degree from Washington University and is now a professor of molecular biology and genetics and senior investigator of the Howard Hughes Medical Institute at the Johns Hopkins University. He is a member of the National Academy of Sciences and the recipient of a Nobel Prize in medicine. His research is focused on genes involved in cell proliferation.

**MAYNARD OLSON** earned his Ph.D. in chemistry at Stanford University. He is currently a professor of genetics at the Washington University School of Medicine, where he does research on the structure and function of eukaryotic genes.

**STUART ORKIN** received his M.D. degree from the Harvard University School of Medicine, where he is now the Leland Fikes Professor of Pediatric Medicine and an investigator of the Howard Hughes Medical Institute. His major research interests include molecular genetics and the biology of human disease.

**LEON ROSENBERG** received his M.D. degree from the University of Wisconsin. He currently serves as a professor and dean of the medical school at the Yale University School of Medicine. He is a member of the National Academy of Sciences and the Institute of Medicine. His research on medical genetics focuses on membrane function, mitochondrial enzymes, and inherited disorders of amino acid metabolism.

**FRANCIS RUDDLE** received his Ph.D. degree from the University of California, Berkeley. He is now a professor of biology and human genetics at Yale University. He is a member of the National Academy of Sciences and conducts research on somatic cell genetics and differentiation.

**SHIRLEY TILGHMAN** earned her Ph.D. degree in biochemistry from Temple University and is now a professor of life sciences at Princeton University. Her own research interests include mammalian molecular genetics.

**JOHN TOOZE** received his Ph.D. degree from London University. He is now the executive secretary of the European Molecular Biology Organization. He conducts research on the cell and molecular biology of secretion at the European Molecular Biology Laboratory in Heidelberg.

**JAMES WATSON** holds a Ph.D. degree from Indiana University and numerous honorary degrees. He is director of the Cold Spring Harbor Laboratory. He received the Lasker Prize and the Nobel Prize in medicine. He is a member of the National Academy of Sciences, the American Academy of Arts and Sciences, and a foreign member of the Royal Society (London).

APPENDIX

# C
# Invited Speakers at Committee Meetings

HOWARD BILOFSKY, Bolt Beranek and Newman Inc.
GEORGE CAHILL, Howard Hughes Medical Institute
ELLSON CHEN, Genentech
ROBERT COOK-DEEGAN, Office of Technology Assessment
GEORGE CHURCH, Harvard University
KAY E. DAVIES, University of Oxford
RONALD W. DAVIS, Stanford University School of Medicine
HELEN DONIS-KELLER, Collaborative Research, Inc.
ARGIRIS EFSTRADIATIS, Columbia University College of Physicians and Surgeons
DAVID GEORGE, National Biomedical Research Foundation Georgetown University Medical Center
JAMES GUSELLA, Massachusetts General Hospital
PATRICIA HOBEN, Office of Technology Assessment
RUTH KIRSCHSTEIN, National Institute of General Medical Sciences/National Institutes of Health
ERIC LANDER, Whitehead Institute for Biological Research and Harvard University
DANIEL MASYS, National Library of Medicine/National Institutes of Health
DAVID PATTERSON, Eleanor Roosevelt Institute for Cancer Research, Inc.
DAVID SMITH, Health Effects Research Division, U.S. Department of Energy

ALAN SPRADLING, Carnegie Institute
JEAN WEISSENBACH, Institute Pasteur
RAY WHITE, Howard Hughes Medical Institute, University of Utah
JOHN C. WOOLEY, Biological Instrumentation Program, National Science Foundation
JAMES WYNGAARDEN, National Institutes of Health

# Index